SpringerBriefs in Molecular Science

SpringerBriefs in Molecular Science present concise summaries of cutting-edge research and practical applications across a wide spectrum of fields centered around chemistry. Featuring compact volumes of 50 to 125 pages, the series covers a range of content from professional to academic. Typical topics might include:

- A timely report of state-of-the-art analytical techniques
- A bridge between new research results, as published in journal articles, and a contextual literature review
- A snapshot of a hot or emerging topic
- An in-depth case study
- A presentation of core concepts that students must understand in order to make independent contributions

Briefs allow authors to present their ideas and readers to absorb them with minimal time investment. Briefs will be published as part of Springer's eBook collection, with millions of users worldwide. In addition, Briefs will be available for individual print and electronic purchase. Briefs are characterized by fast, global electronic dissemination, standard publishing contracts, easy-to-use manuscript preparation and formatting guidelines, and expedited production schedules. Both solicited and unsolicited manuscripts are considered for publication in this series.

Kunal Roy • Arkaprava Banerjee

Activity Cliffs

Where QSAR Predictions Fail

 Springer

Kunal Roy ⓘD
Drug Theoretics and Cheminformatics
Laboratory, Department of Pharmaceutical
Technology
Jadavpur University
Kolkata, West Bengal, India

Arkaprava Banerjee ⓘD
Drug Theoretics and Cheminformatics
Laboratory, Department of Pharmaceutical
Technology
Jadavpur University
Kolkata, West Bengal, India

ISSN 2191-5407   ISSN 2191-5415 (electronic)
SpringerBriefs in Molecular Science
ISBN 978-3-032-10080-1   ISBN 978-3-032-10081-8 (eBook)
https://doi.org/10.1007/978-3-032-10081-8

Foreword

Over the last 50 years, we have witnessed numerous information revolutions. Apparently, the development of predictive computational models in chemistry and biology has transformed modern drug discovery, materials science, and toxicological assessments. Among these methodologies, quantitative structure-activity relationships (QSAR) have established themselves as one of the most powerful and versatile tools of predictive cheminformatics. Nowadays, the drug-discovery toolbox is overflowing with algorithms, datasets, and confidence. We build quantitative structure-activity relationship models, feed them chemical descriptors and matrices of bioactivity, and expect them to sort promising molecules from flops. However, as science advances and datasets become increasingly large and complex, the limitations of QSAR models become equally evident. More often than we like to admit, those expectations are dashed by a deceptively simple phenomenon: an apparently trivial structural change produces a dramatic, unexpected change in activity. These abrupt, local discontinuities in structure-activity relationships are called **activity cliffs**, and they are where QSAR predictions most frequently fail.

This book, *Activity Cliffs: Where QSAR Predictions Fail*, written by leading experts in cheminformatics, explores this phenomenon in depth. It aims to illuminate not only the conceptual underpinnings of activity cliffs but also their implications for predictive modeling, data interpretability, and chemical space exploration. By doing so, it brings attention to one of the most persistent challenges in QSAR modeling, while also pointing toward methodological innovations designed to address it. There is no single magic descriptor or model that will eliminate activity cliffs from your datasets. Instead, this book guides you by offering a pragmatic toolkit: methods to *find* cliffs in your data, interpret them using chemical and structural reasoning, and adapt model-building strategies to be robust or at least transparent about their limitations.

The first two chapters provide a comprehensive grounding in QSAR, its principles, applications, and good practices in model development and validation. Special emphasis is given to dataset modelability, applicability domains, and the critical issue of outliers, all of which form the conceptual framework within which activity cliffs must be understood.

The third chapter introduces the idea of activity cliffs in the biological activity landscape, describing their types, methods of detection, and influence on QSAR predictions. This discussion highlights why cliffs remain both a curse and a blessing—an obstacle to reliable prediction, yet also a window into the underlying molecular mechanisms.

The fourth chapter presents the Arithmetic Residuals in K-Groups Analysis (ARKA), a supervised dimensionality reduction algorithm specifically developed for detecting activity cliffs. Through detailed case studies and applications, ARKA is demonstrated to be a robust and flexible approach, not only for identifying cliffs but also for improving classification and regression modeling, particularly with small or complex datasets.

Finally, this book concludes by situating activity cliffs within the broader discourse on dataset modelability and predictive science. It reflects on how the continuing investigation of activity cliffs can inform future QSAR methodology, enhance interpretability, and ultimately improve the reliability of computational predictions in chemistry and biology.

A recurring theme is humility: good modeling begins with asking the right questions of your data and accepting that predictions must be accompanied by calibrated uncertainty. Where models fail, they teach us about the chemistry we do not yet capture. Activity cliffs are not merely nuisances—they are also opportunities to uncover mechanistic insights and improve both experiments and models.

This volume was written to provide researchers and students with a clear conceptual and methodological guide for understanding activity cliffs and to inspire new research directions that can transform this challenge into an opportunity. As such, this book will serve as both a reference and a stimulus, encouraging further exploration of the boundaries where QSAR predictions fail—and where, perhaps, the most interesting science begins.

Jackson State University
Jackson, MS, USA
Jerzy Leszczynski,

Preface

Quantitative structure-activity relationship (QSAR) and related statistical modeling techniques are routinely used in data gap filling for chemicals used in medicinal chemistry, materials science, predictive toxicology, nanoscience, food science, agricultural science, etc. To develop acceptable QSAR models, the best practices should be followed in light of the guidelines recommended by the Organisation for Economic Co-operation and Development (OECD). It is also important to check the modelability of a QSAR dataset, as the quality of the final models greatly depends on several dataset-specific factors.

Identifying outliers is very important, as they may indicate important physico-chemical features not yet considered during model development and may be a starting point for exploring a different mechanism of action. Some of the prediction outliers may, in reality, be activity cliffs, which exhibit a large difference in activity values between compounds that are quite similar in structural characteristics. The relevance of a QSAR model mainly depends on the chemical domain of the training compounds from which the model has learnt the structure-activity relationships.

The structure-activity landscape of chemical compounds may not always be smooth, though this is a basic assumption of QSAR modeling. Different methods for identifying activity cliffs have been proposed, including the structure-activity landscape index (SALI), the structure-activity relationship index (SARI), and structure-activity similarity (SAS) maps. The Arithmetic Residuals in K-groups Analysis (ARKA) framework has recently been proposed as a supervised dimensionality reduction technique, which is useful for identifying activity cliffs. More recently, a multiclass ARKA framework has been proposed, considering the variable contributions of different QSAR descriptors to different activity ranges. The last approach has been helpful in regression-based model development, particularly when integrated with the similarity-based quantitative read-across structure-activity relationship (q-RASAR) modeling. The similarity levels among different compounds are based on the considered feature set and also the measures used to define the similarity. A comprehensive analysis of the effects of features and similarity definitions on identifying activity cliffs and the resultant impact on the dataset modelability is warranted.

This book introduces readers to the concept of cliffs in predictive cheminformatics, which can significantly impact the modelability of datasets and the quality of predictions, thereby generating disappointment in the performance of QSAR models. With discussions on the standard methods of the detection of applicability domain, analysis of modelability of datasets, and identification of activity cliffs, this book focuses on the recently introduced ARKA approach, which is successful in identifying activity cliffs in several case studies. The application of the multiclass ARKA approach in regression-based problems, integrating it with the q-RASAR framework, has also been covered. Further developments in the ARKA approach may be tracked from https://sites.google.com/site/kunalroyindia/home/arka. This book showcases the evolution and current status of the concept of activity cliffs as relevant to QSAR predictions and indicates future directions in research on activity cliffs. Researchers in the fields of medicinal chemistry, predictive toxicology, nanosciences, food science, agricultural sciences, and materials informatics should benefit from the concept of activity cliffs impacting model-derived predictions.

Kolkata, West Bengal, India   Kunal Roy
Kolkata, West Bengal, India   Arkaprava Banerjee
September 10, 2025

Competing Interests   The authors have no competing interests to declare that are relevant to the content of this manuscript.

Contents

About the Authors

Kunal Roy, PhD, FRSC is Professor and Former Head in the Department of Pharmaceutical Technology, Jadavpur University, Kolkata, India (https://sites.google.com/site/kunalroyindia). He is a recipient of the *Commonwealth Academic Staff Fellowship* (University of Manchester, 2007) and the *Marie Curie International Incoming Fellowship* (University of Manchester, 2013) and a former visiting scientist at the Istituto di Ricerche Farmacologiche "Mario Negri" IRCCS, Milano, Italy. The field of his research interest is quantitative structure-activity relationship (QSAR) and molecular modeling with application in drug design, property modeling, and predictive ecotoxicology. Dr. Roy has published more than 450 research articles (ORCID: http://orcid.org/0000-0003-4486-8074) in refereed journals (current SCOPUS h index 57; total citations to date more than 17,500). He has also coauthored 3 QSAR-related books (Academic Press and Springer), edited 13 QSAR books (Springer, Academic Press, and IGI Global), and published 25 book chapters. Dr. Roy is the Co-Editor-in-Chief of *Molecular Diversity* (Springer Nature) and an Associate Editor of the *Computational and Structural Biotechnology Journal* (Elsevier). Dr. Roy serves on the Editorial Boards of several international journals, including (1) *European Journal of Medicinal Chemistry* (Elsevier); (2) *Journal of Molecular Graphics and Modelling* (Elsevier); (3) *Chemical Biology and Drug Design* (*Wiley*); and (4) *Expert Opinion on Drug Discovery* (Informa). Apart from this, Prof. Roy is a regular reviewer for QSAR papers in the journals, such as *Chemosphere* (Elsevier), *Journal of Hazardous Materials* (Elsevier), *Ecotoxicology and Environmental Safety* (Elsevier), *Journal of Chemical Information and Modeling* (ACS), *ACS Omega* (ACS), *RSC Advances* (RSC), *Molecular Informatics* (Wiley), and *SAR and QSAR in Environmental Research* (T&F). Prof. Roy has been recipient of several awards, including the AICTE Career Award (2003–2004), DST Fast Track Scheme for Young Scientists (2005), multiple Bioorganic and Medicinal Chemistry Most Cited Paper Awards (2003–2006, 2004–2007, and 2006–2009) from Elsevier, The Netherlands, Bioorganic and Medicinal Chemistry Letters Most Cited Paper Award (2006–2009) from Elsevier, The Netherlands, and the Professor R. D. Desai 80th Birthday Commemoration Medal & Prize (2017) from the Indian Chemical Society. Prof. Roy has been a participant in the EU-funded projects nanoBRIDGES and IONTOX, as well as in several national government-funded projects (UGC, AICTE, CSIR, ICMR, DBT, and DAE). Prof. Roy was recently placed in the list of the World's Top 2% science-wide author database (whole career data), achieving a World Rank of 52 in the subfield of Medicinal and Biomolecular Chemistry (Ioannidis, John P.A. (2025), "August 2025 data-update for Updated science-wide author databases of standardized citation indicators", Elsevier Data Repository, V8, link: https://doi.org/10.17632/btchxktzyw.8).

Arkaprava Banerjee, MRSC is a Researcher (funded by the Life Sciences Research Board, DRDO, Govt. of India) working at the Drug Theoretics and Cheminformatics Laboratory, Department of Pharmaceutical Technology, Jadavpur University, Kolkata. Mr. Banerjee has 29 research articles published in reputed journals and 4 book chapters, with overall citations of 763 and an h-index of 17 (Scopus). His ORCID identifier is http://orcid.org/0000-0001-8468-0784 . His expertise lies in similarity-based cheminformatic approaches like Read-Across and Read-Across Structure-Activity Relationship (RASAR)—a novel method that combines the concept of QSAR and Read-Across. Mr. Banerjee is also a Java programmer, who has developed various cheminformatic tools based on QSAR, Read-Across, and RASAR; these tools are freely available from the DTC Laboratory Supplementary website. He received the Professor Anupam Sengupta Bronze Medal from Jadavpur University for securing the highest marks in Pharmaceutical Chemistry in the M.Pharm. examination. He has also received a special diploma awarded by the Institute of Biomedical Chemistry, Moscow, Russia (2022), and the ASCCT Travel Award from the American Society for Cellular and Computational Toxicology (2023). Together with Prof. Kunal Roy, he has been one of the first researchers to develop quantitative models using similarity and error-based descriptors (quantitative/classification Read-Across Structure-Activity Relationship: q-RASAR/c-RASAR models), with applications in drug design, materials science, and property modeling. Recently, he coauthored a book on "q-RASAR," which was published by Springer. He has also co-edited three volumes of *Materials Informatics*, published by Springer. Recently, he was placed in the list of the World's Top 2% science-wide author database (single-year data 2024), achieving a World Rank of 769 in the subfield of Toxicology (Ioannidis, John P.A. (2025), "August 2025 data-update for Updated science-wide author databases of standardized citation indicators", Elsevier Data Repository, V8, link: https://doi.org/10.17532/btchxktzyw.8).

Chapter 1
QSAR: A Tool of Predictive Cheminformatics

Abstract Quantitative structure-activity relationship (QSAR) and related statistical modeling techniques (such as Quantitative structure-property/toxicity relationship or QSPR/QSTR) are routinely used in data gap filling for chemical compounds for which the experimental response (activity/property/toxicity) values are unavailable. Such predictions have shown useful applications in drug design, materials science, predictive toxicology, nanoscience, food science, and agricultural science, among others. However, the quality of predictions from such models is dependent on various factors, including the quality of the modeling data set (e.g., experimental error, data set size, chemical diversity, distribution of response values, presence of influential observations, etc.) and the modeling workflow (e.g., curation and splitting of the data set, variable selection, internal and external validation, consideration of applicability domain, etc.) utilized to develop and validate the model. To develop acceptable QSAR models, the best practices should be followed in light of the guidelines recommended by the Organisation for Economic Co-operation and Development (OECD). It is also important to identify the modelability of a QSAR data set, as the quality of the final models greatly depends on several dataset-specific factors.

Keywords QSAR · Validation · OECD · Modelability · Chemical diversity

1.1 Introduction to QSAR

The exponential growth of Science and Technology in the present times has brought about various discoveries and inventions that benefit mankind. This has been made possible by the contributions of various researchers worldwide who continually work on exploring and developing new techniques. In the early 2000s and earlier times, priority was given to the experimental testing of molecules for the identification of their potential pharmacological actions, toxicity across various living species, and their chemical properties. However, in the last couple of decades, there has been more emphasis on using "New Approach Methodologies (NAMs)" that can

K. Roy, A. Banerjee, *Activity Cliffs*, SpringerBriefs in Molecular Science, https://doi.org/10.1007/978-3-032-10081-8_1

potentially prioritize molecules for experimental testing, considering that such tests are time-consuming, resource-intensive, and attract ethical concerns [1]. The recent press release by the United States Food and Drug Administration (US FDA), dated April 10, 2025 summarizes that the FDA is looking to replace animal testing with NAMs (https://www.fda.gov/news-events/press-announcements/fda-announces-plan-phase-out-animal-testing-requirement-monoclonal-antibodies-and-other-drugs). Additionally, on April 29, 2025, the National Institute of Health (NIH) announced that they will prioritize "human-based research technologies" while reducing animal use in research (https://www.nih.gov/news-events/news-releases/nih-prioritize-human-based-research-technologies), thus potentially accelerating non-animal research models. These announcements will further boost the adoption of NAMs.

One of the most widely used NAMs is the development of computational models. As early as in the 1960s, Corwin Hansch and Toshio Fujita felt the need for establishing mathematical relationships between the structural and physicochemical properties of molecules and the experimentally-derived activity values [2, 3]. This laid to the foundation of "Quantitative Structure-Activity Relationship (QSAR)" models, where one can "predict" the target activity by simply using a mathematical equation. Typically, the simplest form of a QSAR model is a multivariate linear regression equation where the coefficients represent the contributions of structural features of molecules towards the response, i.e., biological activity (Eq. 1.1) [4].

$$Y = a_0 + a_1 \times x_1 + a_2 \times x_2 + a_3 \times x_3 + \ldots a_n \times x_n \tag{1.1}$$

Here, Y is the target/biological activity, $x_1 \ldots . . x_n$ are the essential structural and physicochemical features (a.k.a. descriptors) of molecules, and a_0 is a constant term. Therefore, when we know the values of the corresponding descriptors for a molecule, we can easily calculate its biological activity.

The real challenge for a QSAR modeler is to develop a "robust" QSAR model. This involves training the model on a set of data consisting of known experimental biological activity. Ideally, this "training" data should have a sufficient number of data points with a significant level of heterogeneity in their molecular structures, aiming to encode a vast chemical space [5]. Once the models are developed, they are then tested on external/"unseen" compounds to assess the predictive ability. Typically, this workflow can be related to real-life cases. We study books and various other literature sources to prepare ourselves for an exam, which can be attributed to the "training" of our mind on a specific subject. On the other hand, the questions we receive in an examination are "unseen" to us, and it depends on our mind's "training" to perfectly answer the unseen questions. More details on the assessment of robustness and predictivity of a QSAR model are discussed later in this chapter.

Thus far, we have discussed the classical concepts of QSAR modeling where we emphasize the need for establishing linear relationships between the structural and physicochemical features of molecules and their biological activity. However, this is not always the case. Recent advances infer that many descriptors may have a typically "non-linear" relationship with the target endpoint, which warrants the integration of various Machine Learning (ML) and Deep Learning (DL) modeling approaches within QSAR [6]. The application of these algorithms resulted in improved model performances in many cases; however, these approaches typically compromise with the "mechanistic interpretability" as these ML and DL algorithms are mostly black-box in nature. This is quite an interesting topic of discussion as some research groups prefer improved predictivity, while regulatory agencies like the OECD and EU-REACH stress more on mechanistic interpretability of models, thereby encouraging the development of simpler models with sufficient predictivity. Therefore, it is evident that "predictive QSARs" are sufficiently different from "mechanistic QSARs", which opens the perspective of the modeler to adopt any of them depending on the purpose of model development [7, 8].

1.2   QSAR and Its Applications in Computational Molecular Science

The simplicity of QSAR models has been the driving force behind their increasing use in various scientific fields. Typically, their adoption is boosted by the fact that QSAR modeling simply needs a computing facility, whereas experimental studies are complicated, expensive, and time-consuming. With the increasing demand for drug discovery and the need to fill toxicity data gaps for chemicals, QSAR modeling has proven to be an invaluable tool. This is evidenced by the recent COVID-19 pandemic, where various research groups worldwide have adopted computational modeling techniques to discover new drugs, repurpose existing drug molecules, and develop vaccines, ultimately aiming to reduce the fatality rate [9–11]. Depending on the field of applications, QSAR modeling can be broadly classified into QSAR (when performed on activity endpoints), QSPR (when performed on property-based endpoints), and QSTR (when performed on toxicity endpoints). In this section, we will discuss some of the applications of QSARs in different fields.

1.2.1   Applications in Drug Discovery and Drug Repurposing

New drug discovery and repurposing of existing drug molecules are extremely important research avenues, considering the need for improved drug molecules with reduced side effects, increased bioavailability, and enhanced potency. Their need is further exacerbated by the increasing antibiotic resistance among various bacterial species—a problem that affects millions worldwide. Additionally, exploring structure-activity relationships (SARs) has resulted in the optimization of lead molecules, which present improved efficacy with reduced side effects. For example, in the case of fluoroquinolones, Lomefloxacin exhibits phototoxicity due to the

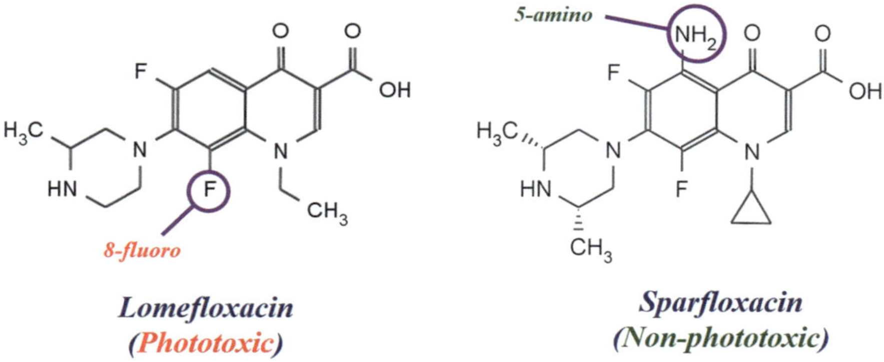

Lomefloxacin
(Phototoxic)

Sparfloxacin
(Non-phototoxic)

Fig. 1.1 Minute structural difference in fluoroquinolones can govern their phototoxicity

presence of a halogen atom at 8-position; however, this is not the case for Sparfloxacin which contains an amino group at 5-position (Fig. 1.1) [12].

Several researchers have worked on developing computational models aiming to predict activity values of drug and drug-like molecules. Among representative examples, we may mention QSAR models developed by Alex Tropsha's group to predict potentially novel antimalarial compounds [13]. In this work, an external set screening prioritized a set of compounds as potential novel antimalarials, and a fraction of those hits was experimentally found to have antimalarial effects. Similarly, Cordeiro's group developed a multi-target QSAR model to predict the anti-breast cancer potency of drugs [14]. The model showed appreciable correct classifications percentages for the actives and inactives. Moreover, the authors analyzed various structural features that are responsible for the anti-breast cancer activity. In another work, Rath et al. [15] trained QSAR models that predict various pharmacokinetic properties. These models were integrated into an expert system where the user can simply input the molecular structure(s) to generate the predictions. Roy's group [16] developed both QSAR and classification Read-Across Structure-Activity Relationship (c-RASAR) models to predict the blood-brain barrier permeability of molecules efficiently. Their models show significantly high prediction accuracies across training and test sets, and demonstrate very good generalizability on a true external set of data.

1.2.2   Applications in Predictive Toxicology

In the field of toxicology, reliable prediction data are invaluable! As of now, very little is known about the toxicity of various kinds of organic and inorganic chemicals across various living species, resulting in a significant data-gap. As stated previously, the experimental determination of an array of different toxicological endpoints for different compounds is time-consuming, resource-intensive, ethically concerning, and inefficient in coping with current demands. Fast, inexpensive, and non-animal approaches are therefore the need of the hour. The use of computational models and other NAMs is currently being encouraged by bodies like the Organisation for Economic Co-operation and Development (OECD), NIH, and EPA. Additionally, the European Commission (EC) has banned the use of animal experimentation on cosmetic products (https://single-market-economy.ec.europa. eu/sectors/cosmetics/ban-animal-testing_en). QSAR models can thus be used as efficient tools for quickly and easily filling the toxicological data gaps, thereby enhancing chemical safety.

Several research groups applied QSAR modeling in the field of predictive (eco) toxicology. Benfenati's group developed QSAR models for the prediction of acute oral toxicity of pesticides against Bobwhite quail [17]. Their model achieved good accuracy in predictions for the training and external validation sets. Roy's group developed QSAR and quantitative Read-Across Structure-Activity Relationship (q-RASAR) models to efficiently predict the rat androgen receptor binding affinity of endocrine disruptors [18]. Their models demonstrated good robustness and improved predictivity, as compared to the previously reported models. Gramatica's group developed QSAR models to predict the acute toxicity of Personal Care Products (PCPs) against algae, *Daphnia*, and fish [19]. The developed models showed good robustness and predictivity even at multiple combinations of training and test set compounds. Cronin's group in 2001 performed QSAR modeling to predict the toxicity of aromatic compounds against *Tetrahymena pyriformis* [20]. Their models showed reasonable explained variances across the different modeling methodologies.

Apart from the conventional QSAR modeling, other similarity-based data gap filling approaches like Read-Across and q-RASAR have found increasing use and acceptability in predictive toxicology setting [21, 22]; however, their application is not limited to predicting toxicological endpoints, but also extended to drug discovery and materials science [16, 23].

1.2.3 Applications in Materials Science and Material Property-Based Endpoints

QSAR modeling has been a very useful tool in the field of materials science. These models can efficiently predict materials properties that can be helpful to optimize materials design. Gramatica's group developed QSPR models to predict the soil sorption partition coefficient ($log\ K_{OC}$) of compounds [24]. The various internal and external validation metrics suggest that the developed models were highly robust and predictive. Alan Katritzky and Mati Karelson's group, in 1998, developed QSPR models to predict the glass transition temperature (T_g) of high-molecular-weight polymers [25]. Their models showed high fitting statistics. Later, Roy's group also worked on developing QSPR models for predicting T_g of diverse polymers [26]. Their model also showed good fitting and predictive ability. Rojas et al. [27] developed QSPR models to predict the retention index of flavors and fragrances in a gas chromatography column. The modeling results suggest that their model not only had good fitting ability, but also showed promising external predictive ability.

A representation of the applications of QSARs in different fields has been depicted in Fig. 1.2, and a summary of the different applications has been presented in Table 1.1.

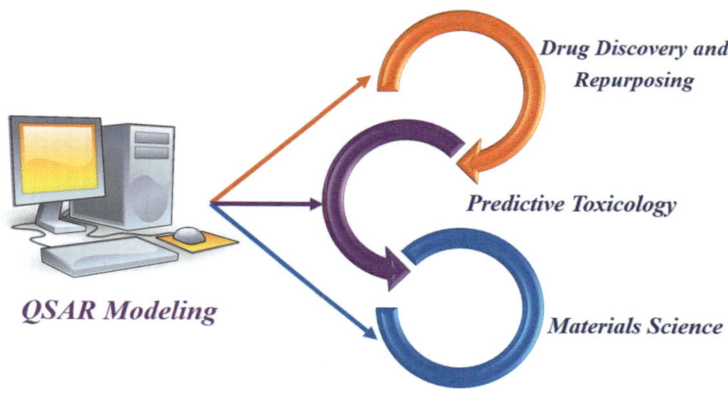

Fig. 1.2 Applications of QSARs in different fields

Table 1.1 Summary of some of the applications of QSARs in different scientific domains

Research groups	Ref. no.	DOI
Drug discovery and drug repurposing		
Zhang et al.	[13]	https://doi.org/10.1021/ci300421n
Speck-Planche et al.	[14]	https://doi.org/10.1016/j.ejps.2012.04.012
Rath et al.	[15]	https://doi.org/10.1021/acs.jmedchem.3c02446
Kumar et al.	[16]	https://doi.org/10.1021/acs.jcim.4c00433
Predictive toxicology		
Iovine et al.	[17]	https://doi.org/10.3390/environments12020056
Banerjee et al.	[18]	https://doi.org/10.1016/j.chemosphere.2022.136579
Gramatica et al.	[19]	https://doi.org/10.1039/C5GC02818C
Cronin and Schultz	[20]	https://doi.org/10.1021/tx0155202
Materials science and materials property-based endpoints		
Gramatica et al.	[24]	https://doi.org/10.1016/j.jmgm.2006.06.005
Katritzky et al.	[25]	https://doi.org/10.1021/ci9700687
Khan and Roy	[26]	https://doi.org/10.1080/1062936X.2018.1536078
Rojas et al.	[27]	https://doi.org/10.1016/j.chemolab.2014.09.020

1.3   Good Practices of QSAR Model Development and Validation

Paola Gramatica's group, in one of their papers, inferred that QSAR modeling is not "push a button and find a correlation" [28]. As QSAR modelers, it is important to adhere to the "Best Practices" of QSAR modeling to enhance the acceptability and reliability of the developed models [29]. These practices include sourcing reliable biological data, curation of the input data, identifying outliers (i.e., compounds beyond the feature space of the developed model), eliminating multi-collinearity among descriptors, avoiding model overfitting, capturing non-linear information, computing various internal and external validation metrics, proper reporting of data for the ease of reproducibility and transferability, and proposing a mechanistic interpretation of the contributing features [5]. This section provides an in-depth analysis of these aspects.

1.3.1   Sourcing Reliable Biological Data

One of the primary and most important aspects of developing QSAR models is to collect a "reliable" dataset for model development. This is very essential, as it dictates the quality and reliability of the developed model. Ideally, the experimental response data (to be modelled) should be of the same unit, measured against the same target animal species in the same laboratory and environmental conditions, using well-standardized assays. The collected data set should contain a sufficient number of data points. Additionally, each biological experiment contains errors, and therefore, it is essential to assess their reliability by considering the dispersion of the experimental data for the same data point [5]. If the differences in values are statistically insignificant, we can infer that the data is of high quality.

1.3.2   Chemical Curation of the Molecular Structures

This is another important aspect to consider before proceeding towards model development. Each chemical structure should be checked and analyzed for the presence of mixtures, salt forms, and duplicates. Additionally, depending on the complexity level of chemical information used to develop the QSAR model, the nature of "duplicates" may change. For example, when we are interested in developing 3D-QSAR models, a pair of stereoisomers is treated as two distinct compounds, as the model effectively captures the spatial orientation of the atoms. However, if the aim is to develop a 2D-QSAR model, which typically does not encode information related to the spatial orientation of atoms, a pair of stereoisomers may be treated as "duplicates" and one of them must be removed from the dataset. This is important because retaining both may eventually lead to their distribution in training and test sets. Additionally, normalization of the chemical structures ensures that no compounds are present as zwitterions, which would be potential duplicates if their non-zwitterionic components are already present as separate data points. Therefore, an efficient QSAR modeler should look for these aspects, identify and treat mixtures (either by retaining the largest fragment or removing the data point as a whole), remove the salt components, normalize the chemical structures, and eliminate duplicates. However, manual curation may be error-prone, which is why researchers have developed automated curation workflows that are "QSAR-ready" [30].

1.3.3 Dealing with Multi-collinearity Among Descriptors

Multi-collinearity refers to a situation in which two or more descriptors share a significantly high intercorrelation. This is an important aspect when developing statistical QSAR models. These intercorrelated descriptors tend to multiply the noise in the modeled feature space, thereby generating inflated model statistics with poor generalizability on unseen data [5, 31]. Typically, this contributes to the "curse of dimensionality", where the model learns the "noise" of the associated feature space, thereby generating inflated training set performance with poor prediction performance on the test/unseen data [32]. This can easily be tested by measures such as cross-validation and ablation studies, which have been discussed in detail later in this chapter. As part of the good practices of QSAR model development, the intercorrelated descriptors and the descriptors suffering from low variances across the entire dataset should be removed from the modeling analysis.

1.3.4 Identification of Outliers and Activity Cliffs

Each model, trained with a set of descriptors, encodes a certain feature space. This is directly governed by the molecular composition of the training set. Therefore, for training a reliable QSAR model that encodes a large chemical space, there should be sufficient structural heterogeneity among the training data points. If the training data is too homogenous, this can result in the development of models with a limited feature space, and the predictions for the unseen compounds would be unreliable. For example, if we develop a model to predict the anti-depressant activity of drug molecules, and we use a training set consisting of only tricyclic antidepressants (TCA) like imipramine, it is most likely that other (structurally different) antidepressant drugs like Fluoxetine would generate unreliable predictions. Such compounds may be referred to as compounds that "do not belong to the feature space" of the developed model, and are typically termed outliers. Predictions for such compounds are considered unreliable, irrespective of their outcomes. Therefore, to develop efficient QSAR models, the modeler should ensure sufficient structural diversity in the training set, which enhances the reliability of unseen predictions, and reduces the number of outliers.

Activity cliffs, on the other hand, are pairs of compounds that share a high level of structural similarity with each other; however, they contrastingly differ in their experimental activity. Again, these are governed by the available feature space using which we are "judging" their similarity; missing out on other important features that could have been deterministic to predict their activities correctly. As evident from Sects. 1.3.3 and 1.3.4, it becomes highly crucial to select just the right number and nature of descriptors to develop effective QSAR models. Further details on activity cliffs are given in Chap. 3.

1.3.5   Validating the QSAR Model to Measure the Goodness-of-Fit, Robustness, and Predictivity

Validation of a QSAR model typically refers to testing the model's robustness, fitting capability, and the ability to generalize well on unseen data. For quantitative endpoint data, the model's fitting performance on the training set is estimated by using various correlation and error-based metrics like the determination coefficient (R^2), and the mean absolute error of the training set fitting (MAE_{Train}), respectively. The higher the R^2 and the lower the MAE_{Train} values, the better the model's fitting ability on the training set data. Robustness, on the other hand, represents the model's stability. This is tested by removing a fragment of the training set compounds, in different iterations, and the models are re-developed. Ideally, for a robust model, there should not be a "significant" drop in the model performance [33]. This process is referred to as cross-validation of a model, and is discussed in detail in the following section. Cross-validation typically ensures that the model fitting ability is not solely governed by certain "influential" data points [33], rather the model fitting performance should be based on the entire training set chemical space. Assessing the external predictivity of a model involves the exploration of the model's predictive ability on "unseen" data. This is typically performed by generating model-derived predictions for a set of unseen compounds (test set) that have experimental target response values. Different correlation ($Q_{F1}^2, Q_{F2}^2, r_{Test}^2$) and error-based ($MAE_{Test}$ and $RMSE_{Test}$) metrics are computed to check the quality of predictions. Metrics that measure the goodness-of-fit and robustness of models (i.e., related to the estimation of the performance of the training set) are termed internal validation metrics, while metrics that measure the external predictivity of the model (i.e., related to the estimation of the model performance on the test set) are termed external validation metrics. On the other hand, for graded endpoint data, the correlation and error-based metrics (stated above) make little sense. Therefore, classification-based validation metrics like Accuracy, Balanced Accuracy, Area Under the ROC Curve (AUC), Mathews Correlation Coefficient (MCC), and Cohen's kappa (CKappa), are used to check the model performances on the training and test data. Some of the most important regression and classification-based validation metrics have been provided in Table 1.2. While selecting the best model, the modeler should ensure that there is no significant performance difference between the training and test set compounds, and the model should show similar performance in both the

Table 1.2 Descriptions of the different regression and classification-based validation metrics commonly employed in QSAR studies

Metrics	Descriptions	Formula[a]
Regression		
R^2	Determination coefficient—used to judge the model fitting quality	$R^2 = 1 - \dfrac{\sum\left(Y_{obs} - Y_{calc}\right)^2}{\sum\left(Y_{obs} - \bar{Y}_{obs}\right)^2}$
MAE	Mean absolute error in predictions—defines the average error in predictions	$MAE = \dfrac{\sum\lvert Y_{obs} - Y_{calc}\rvert}{n}$
RMSE	Root mean squared error	$\sqrt{\dfrac{\sum\left(Y_{obs} - Y_{calc}\right)^2}{n}}$
Q^2_{CV}	Cross-validated determination coefficient	$Q^2_{CV} = 1 - \dfrac{\sum\left(Y_{obs} - Y_{CVcalc}\right)^2}{\sum\left(Y_{obs} - \bar{Y}_{obs}\right)^2}$
Q^2_{F1}	Expression to calculate the Q^2 for external (test) compounds	$Q^2_{F1} = 1 - \dfrac{\sum\left(Y_{obs(Test)} - Y_{pred(Test)}\right)^2}{\sum\left(Y_{obs(Test)} - \bar{Y}_{Training}\right)^2}$
Q^2_{F2}	Expression to calculate the Q^2 for external (test) compounds	$Q^2_{F2} = 1 - \dfrac{\sum\left(Y_{obs(Test)} - Y_{pred(Test)}\right)^2}{\sum\left(Y_{obs(Test)} - \bar{Y}_{Test}\right)^2}$
Classification		
Sensitivity	Used to evaluate the predictive performance of positive class data	$Sensitivity = \dfrac{TP}{TP + FN}$
Specificity	Used to evaluate the predictive performance of negative class data	$Specificity = \dfrac{TN}{TN + FP}$
Accuracy	Used to estimate the accuracy in predictions	$Acc = \dfrac{TP + TN}{TP + FN + TN + FP}$
Balanced accuracy	Used to estimate the "balanced" performance on both the classes of data	$BA = \dfrac{Sensitivity + Specificity}{2}$
MCC	Mathews correlation coefficient	$MCC = \dfrac{(TP \times TN) - (FP \times FN)}{\sqrt{(TP + FP)\times(TP + FN)\times(TN + FP)\times(TN + FN)}}$
AUC	Area under the ROC curve	

[a] Y = Biological Activity; Y_{obs} = Observed Biological Activity; Y_{calc}/Y_{pred} = Calculated/Predicted Y values for the training and test sets, respectively; $\bar{Y}$ = Mean Y_{obs}; TP = True Positives; TN = True Negatives; FP = False Positives; FN = False Negatives

seen and unseen data. This is known as Kubinyi's paradox [34], which emphasizes the need for checking the predictive quality for unseen compounds, which should be at par with the training set fitting statistics [35].

1.3.6   Avoiding Model Overfitting

Overfitting refers to a case where the model predicts the training data very well, but their predictivity on unseen data is significantly reduced. This is an indicator that the developed model is not robust and has poor predictivity. There can be various reasons why models often tend to become overfitted, which are discussed below.

1.3.6.1   The Curse of Dimensionality: Considering Too Many Features for Model Development

This is a common cause for getting overfitted models. When the training set is presented by a significantly higher number of descriptors, it is imperative that many descriptors may be redundant and inter-correlated. This contributes to the definition of a "noisy" feature space, which misguides the modeling algorithm from identifying meaningful patterns. The algorithm then learns from the noise to generate overfitted models. In these cases, generally the training set fitting performance is quite good, but the performance on the test set, or even during cross-validation, falls significantly. In a study by Preetha et al. [36], the authors developed QSAR models using only seven (7) compounds, while using as high as four (4) or five (5) descriptors. Evidently, the R^2 values are "unrealistically" high, and the curse of dimensionality has reflected on the sharp decrease in the R^2_{adj} values. The models are therefore invalid.

1.3.6.2   Lack of Sufficient Training Data Points

Overfitting of models is often encountered in modeling small datasets. Lack of sufficient training data points reduces the "reliability" of the fitting statistics of the model. In many cases, it has been observed that a selected number of influential data points governs the model fitting. For example, in the works of Abdullayev et al. [37], many QSAR models reported by the authors had a significant difference among their fitting, prediction, and CV statistics, which should ideally not be the case. This problem can be mitigated either by reducing the dimensionality of the input features [38], or by performing predictions using non-statistical approaches like Read-Across [39]. According to the best practices of QSAR model development, the ratio of the number of training data points to the number of modeling descriptors should be as high as possible, typically greater than 5:1 [40].

1.3.6.3 Improper Tuning of the Model Hyperparameters

This is an aspect that is often encountered during the application of various Machine Learning (ML) algorithms to train QSAR models. Although optimization of hyperparameters of ML algorithms is a standard practice, optimization covering a large space of hyperparameters can potentially generate overfitted models [41]. In tree-based algorithms like Random Forests, increasing the depth and number of trees may potentially result in the development of overfitted models. These "deep" trees learn the training data well, but may perform poorly on test data.

Estimation of robustness and detection of potential model overfitting is a key challenge for QSAR modelers. The modeler should logically select the appropriate techniques and validation metrics to judge the model performance. For example, when we develop QSAR models on a large dataset (data points typically in thousands), it is improper to assess the model robustness using the Leave-One-Out Cross-validation technique (LOO-CV). This algorithm removes a compound in each iteration, redevelops the model from the reduced set of compounds, and predicts the biological activity of the removed compound. For a relatively small dataset, this is an acceptable and robust method; however, considering large datasets, there is an insignificant effect of the "removed compound," and the model statistics remain almost the same during CV. This hinders the proper estimation of the robustness of the developed model. Therefore, in such cases, techniques like the k-fold cross-validation (k-fold CV) or repeated k-fold CV [42] are a good choice for the actual estimation of robustness of the model, as this algorithm involves the removal of a significant part of the training set compounds in each iteration [43]. Additionally, the modeler should be careful during model development of highly imbalanced datasets, as this warrants the adoption of a stratified k-fold CV technique [44], which preserves the ratio of actives and inactives during CV.

1.3.7 Capturing Non-linear Information

The concept of classical QSAR is based on the theory that descriptors should have a certain linear relationship with the biological activity. However, several studies have identified non-linear relationships between the descriptors and the target activity. This can be achieved by the integration of various machine learning algorithms into the QSAR modeling framework. Typically, most of the ML models are non-linear in nature, and can potentially identify intricate relationships between the descriptors and the target endpoint. Algorithms like neural networks act like the human brain, where complex non-linear relationships are identified in its hidden layers. Identifying non-linear relationships is therefore essential, as they can improve the model performance [45].

1.3.8   Proper Documentation of Data and Ensuring Reproducibility

This is one of the most important aspects considering the "good practices" of QSAR model development. The developed model should be made available for the use of other researchers. Ensuring reproducibility and transparency is of utmost importance for the proper use of the developed model. Additionally, complete data availability (like the composition of the training and test set compounds, their descriptor values, and the model hyperparameters) for the developed model promotes reproducibility. It is always a good practice to report the model as per the standard QSAR Model Reporting Format (QMRF) (https://one.oecd.org/document/ENV/CBC/MONO(2023)32/ANN1/en/pdf), where every details of the methodological aspects can be obtained.

1.3.9   Mechanistic Interpretation of the Contributing Features

As outlined by the OECD Principle 5, proposing a mechanistic interpretation of the developed model, if possible, enhances the regulatory acceptability. This provides a clear understanding of the relative contribution of various structural and physico-chemical features towards the target endpoint. The contributions of features provide useful information for synthetic chemists towards developing improved molecules against a particular target.

1.3.10   Miscellaneous Factors That Drive the "Good Practices" of QSAR Modeling

Apart from the points already mentioned previously, there are certain other factors that need to be considered. The endpoint values to be modeled should be in an appropriate form. Often, we find that the range of experimental data in a dataset is very wide. Developing models using such data will inevitably generate inflated correlation-based validation metrics, which may misguide the modeler to assume that the model performs well on the training data. However, in such cases, the error-based metrics like MAE and RMSE reflect the actual predictive power, and we often have very high error values. The "good practice" involves applying a logarithmic (or another suitable) transformation to such experimental data, which aims to reduce the response range, especially in the biological domain. Modeling using the log-transformed data, especially in the biological domain, is likely to efficiently reflect the true model fitting and predictivity [4].

Secondly, if the modeler works on toxicity and activity data, response data like LD_{50}, IC_{50} etc., are generally considered. Ideally, these data should be taken in a molar scale. In addition, to develop reliable models, it is essential to understand what the endpoints depict. For example, LD_{50} refers to the dose of a molecule that is lethal to 50% of a sample population. In a general conscience, lower the lethal dose (LD_{50}), more toxic/potent is the molecule. If QSAR models are developed using "log LD_{50}", higher prediction values will typically denote safer molecules—not the toxic ones. Therefore, the "good practice" involves considering negative logarithmic transformation of the LD_{50} values (pLD_{50}), and using this for developing QSAR models. In this setting, the higher the pLD_{50} value, the more toxic the molecule, and vice versa [4].

Developing classification-based QSAR models involves some additional considerations. This approach is generally considered when the quantitative response data are not available. Such datasets typically report the "active or inactive" status of compounds by using binary notations (0 for inactive and 1 for active). The problem associated with the training data, in such cases, is that there is always a higher number of representatives for one particular class (i.e., actives or inactives). If this difference is significant, the dataset can be considered as skewed, and models developed using this training data will always tend to predict an unseen compound to the class that represents the majority of the training data, due to the induced bias in the model development. To mitigate this effect, "good practices" involve balancing the training set data to have nearly equal representation of the active and inactive classes of data [29]. This can be achieved by adopting various undersampling techniques (i.e., reducing the number of majority class data points to balance with the minority class) and oversampling techniques (i.e., generating virtual data points of the minority class to balance with the majority class) [29].

In many cases, when the training dataset is highly imbalanced (e.g., actives: inactives = 10:1), oversampling of the minority class data points may probably result in noisy virtual data points with a very limited degree of heterogeneity, which potentially limits the chemical space of the developed model. On the other hand, undersampling of the majority class data may result in a final training set with a limited number of data points, associated with a significant loss of chemical information. In such cases, where the modeler decides to proceed towards model development using the imbalanced dataset, it becomes very crucial to select the model cross-validation strategy and appropriate metrics judiciously. As mentioned earlier, the stratified k-fold CV, and not the regular k-fold CV, should be the approach for cross-validating the developed model, as it maintains the class distribution in the ablated set of data points in each iteration. Additionally, selecting the appropriate validation metrics is essential to truly identify the performance of the model. In this representative example, where the model is skewed towards predicting unseen compounds as "actives", it is incorrect to judge the model quality using metrics like Sensitivity, as this metric considers only the active data points, and is susceptible to the number of such data points present in the training set. Similarly, in other cases where the model is skewed towards predicting unseen compounds as "inactives", judging based on Specificity is wrong. Therefore, for highly imbalanced sets, the modeler should stress metrics like "Balanced Accuracy", "Matthews Correlation Coefficient (MCC)", and the "Area Under the ROC Curve (AUC)" for judging the predictive ability of the developed model. While Balanced Accuracy is the average of Sensitivity and Specificity values, the MCC and AUC reflect the overall performance of the developed model. In cases where the modeler does not observe significant differences in the values of Sensitivity, Specificity, and Balanced Accuracy, it can only then be inferred that the developed model does not have significant bias.

1.4 Dataset Modelability and Its Measures

Estimating the modelability of a dataset is crucial before proceeding with model development. This provides the QSAR modeler with an idea of how good a model can be developed theoretically, thereby potentially eliminating the need for further investment of time and labour for improving model performance. So far, various research groups have explored the estimation of the modelability of datasets using various approaches. Some of these approaches have been listed in this section.

1.4.1 The MODI Index

In 2014, Tropsha's group introduced the MODelability Index (MODI) to estimate the modelability of a dataset [46]. This is based on a simple theory that compounds whose first nearest neighbor, in terms of Euclidean distance, belongs to the opposite activity class (in a binary classification problem), can be termed activity cliffs. For a dataset to exhibit significant modelability, it is necessary that, in most cases, the first nearest neighbor should belong to the same activity class. The formula of MODI has been depicted in Eq. 1.2.

$$MODI = \frac{1}{K} \sum_{i=1}^{K} \frac{N_i^{same}}{N_i^{total}} \qquad (1.2)$$

In Eq. 1.2, K stands for the number of classes (whose value is typically "2" in a binary classification problem), N_i^{same} is the number of compounds, belonging to the ith class, where the first nearest neighbors have the same activity class, and N_i^{total} represents the total number of compounds in the ith class.

The authors also stated that the predictive power of a model depends on the correct classification rate (QSAR_CCR), a.k.a., the balanced accuracy. Modification of the MODI formula by replacing N_i^{same} to the number of correctly predicted compounds, would yield QSAR_CCR. A QSAR_CCR value greater than 0.7 infers that the model has an acceptable predictive power [29]. After an extensive analysis of over 100 datasets, the authors inferred that MODI values >0.65 reflect that the dataset is modelable, whose QSAR_CCR would be >0.7. The authors also proposed

MODI diversity index and activity cliff index, which can be found elsewhere [47] and also discussed in Chap. 3.

1.4.2   The Rivality Index

In 2018, Ruiz and Gomez-Nieto extended the MODI calculation by incorporating a Rivality Index, aiming to account for the noise or "rivality" induced by the nearest neighbors [48]. Typically, the formula for MODI considers the first nearest neighbor and the corresponding activity status. However, the rivality index is calculated considering the distance to the nearest active neighbor and the nearest inactive neighbor, as shown in Eq. 1.3.

$$RI_i = \frac{d_i^x - d_i^y}{d_i^x + d_i^y} \tag{1.3}$$

In Eq. 1.3, d_i^x represents the distance between a molecule i and its nearest neighbor belonging to the same activity class of i, while d_i^x represents the distance between a molecule i and its nearest neighbor belonging to a different activity class of i. The value of this rivality index typically ranges from -1 to $+1$, where negative values represent that the first nearest neighbor is of the same class as the query compound i, and vice versa. This has been used to re-formulate MODI, which has been represented in Eq. 1.4.

$$MODI = \frac{1}{k} \sum_{j=1}^{k} \left[\frac{1}{M_k} \sum_{i=1}^{M_k} \left(RI_i \leq 0 \right) \right] \tag{1.4}$$

In Eq. 1.4, K is the number of classes, M_k represents the number of molecules belonging to the class K, and RI is the rivality index. The advantage of this modified MODI formula is that it can indicate cases where the distance to the closest neighbor of the same class and the closest neighbor to the other class, for a particular compound, is same.

Additionally, the authors also extended their calculation of rivality index by applying weights to the distances to the first neighbors. This is because the authors observed that it is not always the first neighbors of a molecule that governs their correct classification. Rather, the subsequent neighbors may have the same class distribution, which in turn may or may not be the same for the query molecule. Therefore, in order to consider this effect of the subsequent neighbors, the authors computed weights for the distances to the first neighbors, which in turn can be used to compute the Weighted MODI (WMODI). The weighted RI and the WMODI are mentioned in Eqs. 1.5 and 1.6, where the weights are denoted as w_i^x and w_i^y.

$$\overline{RI}_i = \frac{\left(d_i^{\bar{z}} w_i^x\right) - \left(d_i^y w_i^y\right)}{\left(d_i^z w_i^x\right) + \left(d_i^y w_i^y\right)} \tag{1.5}$$

$$WMODI = \frac{1}{k}\sum_{j=1}^{k}\left[\frac{1}{M_k}\sum_{i=1}^{M_k}(\overline{RI}_i \leq 0)\right] \tag{1.6}$$

1.4.3   Banerjee and Roy's Concordance and Similarity Indices

Thus far, the focus has mostly centered on the effect of the first nearest neighbors of compounds, and their relation to the modelability of a dataset. However, this is not always the case. In many instances, the similarity to the other nearest neighbors is not drastically lower than that to the first nearest neighbors, suggesting that these compounds would also contribute to affecting modelability. Additionally, as also noted by Ruiz and Gomez-Nieto [48], there may be two compounds among the nearest neighbors having the same value of distance/similarity but of opposite class labels, and in such cases, the traditional MODI index has limitations, which is the reason for proposing the weighted RI.

In 2022 and 2023, Banerjee and Roy developed a concordance measure (g_m) and two similarity measures (s_m^1 and s_m^2) [49, 50], considering various factors contributing to modelability. These include the consideration of the similarity to the nearest active and inactive neighbor for a molecule, the fraction of nearest neighboring compounds that are active/inactive, and the average similarities of the nearest active and inactive compounds. The formulae of g_m, s_m^1, and s_m^2 have been depicted in Eqs. 1.7–1.9.

$$g_m = (-1)^n \times 2 \,|\, PosFrac - 0.5 \,| \tag{1.7}$$

$$s_m^1 = \frac{MaxPos - MaxNeg}{argmax\left(MaxPos,\ MaxNeg\right)} \tag{1.8}$$

$$s_m^2 = \frac{PosAvgSim - NegAvgSim}{AvgSim} \tag{1.9}$$

In Eqs. (1.7)–(1.9), *PosFrac* is the fraction of active molecules among the list of close source neighbors, *MaxPos* is the similarity value towards the nearest active neighbor, *MaxNeg* is the similarity value to the nearest inactive neighbor, *PosAvgSim* is the average similarity of only the active nearest neighbors, *NegAvgSim* is the average similarity of only the inactive nearest neighbors, and *AvgSim* is the average similarity of the nearest neighbors irrespective of their class labels. In Eq. 1.7, the value of n is either "1" when $MaxPos < MaxNeg$, or "2" when $MaxPos \geq MaxNeg$. Therefore, the values of g_m are negative when the similarity to the closest inactive neighbor is more than the closest active neighbor, and vice versa. When the value of g_m becomes 0, which typically happens when $PosFrac$ is 0.5 (i.e., the nearest neighbors have equal proportions of active and inactive compounds), it becomes difficult to conclude the modelability. This is why the authors proposed the metric g_m_class [50], whose values are 0 or 1 when g_m values are negative or positive, respectively. However, in cases where $g_m = 0$, it considers whether $MaxPos$ or $MaxNeg$ is higher. In the former case, i.e. where $MaxPos$ is higher, g_m_class becomes 1, and in the latter case, g_m_class becomes 0. Therefore, for a binary classification problem, if the values of g_m_class and the activity labels match in most cases, it may be inferred that the dataset is modelable, and vice versa.

The metrics s_m^1 and s_m^2 also follow a similar principle. For active compounds, we expect a higher similarity value to the closest active neighbor, and a higher average similarity value among the closest active neighbors. This typically should result in positive values of s_m^1 and s_m^2; however, this is not ideally the case. In many cases, active compounds may have positive values for either of s_m^1 or s_m^2, or inactive compounds may have negative values for either of s_m^1 or s_m^2. Additionally, in some cases, there may be a complete opposite observation, like active compounds having negative s_m^1 and s_m^2 and inactive compounds having positive s_m^1 and s_m^2 values. These compounds can be considered activity cliffs, which will be discussed in detail in Chap. 3. Datasets having a greater number of such "opposite observations" can be termed as less modelable datasets.

1.5   Applicability Domain and Modelability: The Importance of Chemical Space

In feature-based modeling approaches, the modelability of a dataset is primarily governed by the selected features, using which the similarities are defined. These features generate a "chemical space" a.k.a. Applicability Domain (AD), where most of the data points are expected to lie. However, there may be certain data points that lie outside this common feature space, and affect the modelability of the dataset. These data points are termed Outliers, where the structural features do not seem to be included within the existing feature space. Predictions for such compounds are generally considered unreliable. An insight into the concepts of Applicability Domain and Outliers has been provided in Chap. 2, where the readers can understand their role in misleading predictions and the necessity of developing a model from a sufficiently large chemical space.

References

1. Ball N, Bars R, Botham PA, Cuciureanu A, Cronin MTD, Doe JE, Dudzina T, Gant TW, Leist M, Ravenzwaay BV (2022) A framework for chemical safety assessment incorporating new approach methodologies within REACH. Arch Toxicol 96:743–766. https://doi.org/10.1007/s00204-021-03215-9
2. Hansch C, Fujita T (1964) p-σ-π analysis. A method for the correlation of biological activity and chemical structure. J Am Chem Soc 86:1616–1626. https://doi.org/10.1021/ja01062a035
3. Hansch C, Maloney PP, Fujita T, Muir RM (1962) Correlation of biological activity of phenoxyacetic acids with hammett substituent constants and partition coefficients. Nature 194:178–180. https://doi.org/10.1038/194178b0
4. Roy K, Kar S, Das RN (2015) Understanding the basics of QSAR for applications in pharmaceutical sciences and risk assessment. Academic Press, New York. https://doi.org/10.1016/C2014-0-00286-9
5. Cronin MTD, Schultz TW (2003) Pitfalls in QSAR. J Mol Struct Theochem 622:39–51. https://doi.org/10.1016/S0166-1280(02)00616-4
6. Tropsha A, Isayev O, Varnek A, Schneider G, Cherkasov A (2024) Integrating QSAR modelling and deep learning in drug discovery: the emergence of deep QSAR. Nat Rev Drug Discov 23:141–155. https://doi.org/10.1038/s41573-023-00832-0
7. Bhhatarai B, Garg R, Gramatica P (2010) Are mechanistic and statistical QSAR approaches really different? MLR studies on 158 cycloalkyl-pyranones. Mol Inform 29:511–522. https://doi.org/10.1002/minf.201000011
8. Fujita T, Winkler D (2016) Understanding the roles of the "two QSARs". J Chem Inf Model 56:269–274. https://doi.org/10.1021/acs.jcim.5b00229
9. Kumar V, Roy K (2020) Development of a simple, interpretable and easily transferable QSAR model for quick screening antiviral databases in search of novel 3C-like protease (3CLpro) enzyme inhibitors against SARS-CoV diseases. SAR QSAR Environ Res 31:511–526. https://doi.org/10.1080/1062936X.2020.1776388
10. Kumar V, Kar S, De P, Roy K, Leszczynski J (2022) Identification of potential antivirals against 3CLpro enzyme for the treatment of SARS-CoV-2: a multi-step virtual screening study. SAR QSAR Environ Res 33:357–386. https://doi.org/10.1080/1062936X.2022.2055140
11. Alves VM, Bobrowski T, Melo-Filho CC, Korn D, Auerbach S, Schmitt C, Muratov E, Tropsha A (2021) QSAR modeling of SARS-CoV Mpro inhibitors identifies sufugolix, cenicriviroc, proglumetacin, and other drugs as candidates for repurposing against SARS-CoV-2. Mol Inform 40:2000113. https://doi.org/10.1002/minf.202000113
12. Beale Jr JM, Block JH (2011) Wilson and Gisvold's textbook of organic medicinal and pharmaceutical chemistry.12th edn. Lippincott Williams & Wilkins, Philadelphia
13. Zhang L, Fourches D, Sedykh A, Zhu H, Golbraikh A, Ekins S, Clark J, Connelly MC, Sigal M, Hodges D, Guiguemde A, Guy RK, Tropsha A (2013) Discovery of novel antimalarial compounds enabled by QSAR-based virtual screening. J Chem Inf Model 53:475–492. https://doi.org/10.1021/ci300421n
14. Speck-Planche A, Kleandrova VV, Luan F, Cordeiro MNDS (2012) Chemoinformatics in anti-cancer chemotherapy: multi-target QSAR model for the in silico discovery of anti-breast cancer agents. Eur J Pharm Sci 47:273–279. https://doi.org/10.1016/j.ejps.2012.04.012
15. Rath M, Wellnitz J, Martin HJ, Melo-Filho C, Hochuli JE, Silva GM, Beasley JM, Travis M, Sessions ZL, Popov KI, Zakharov AV, Cherkasov A, Alves V, Muratov EN, Tropsha A (2024) Pharmacokinetics profiler (PhaKinPro): model development, validation, and implementation as a web tool for triaging compounds with undesired pharmacokinetics profiles. J Med Chem 67:6508–6518. https://doi.org/10.1021/acs.jmedchem.3c02446

16. Kumar V, Banerjee A, Roy K (2024) Breaking the barriers: machine-learning-based c-RASAR approach for accurate blood–brain barrier permeability prediction. J Chem Inf Model 64:4298–4309. https://doi.org/10.1021/acs.jcim.4c00433
17. Iovine N, Roncaglioni A, Benfenati E (2025) Predicting acute oral toxicity in bobwhite quail: development of QSAR models for LD50. Environments 12:56. https://doi.org/10.3390/environments12020056
18. Banerjee A, De P, Kumar V, Kar S, Roy K (2022) Quick and efficient quantitative predictions of androgen receptor binding affinity for screening endocrine disruptor chemicals using 2D-QSAR and chemical read-across. Chemosphere 309:136579. https://doi.org/10.1016/j.chemosphere.2022.136579
19. Gramatica P, Cassani S, Sangion A (2016) Aquatic ecotoxicity of personal care products: QSAR models and ranking for prioritization and safer alternatives' design. Green Chem 18:4393–4406. https://doi.org/10.1039/C5GC02818C
20. Cronin MTD, Schultz TW (2001) Development of quantitative structure−activity relationships for the toxicity of aromatic compounds to Tetrahymena pyriformis: comparative assessment of the methodologies. Chem Res Toxicol 14:1284–1295. https://doi.org/10.1021/tx0155202
21. Banerjee A, Kar S, Roy K, Patlewicz G, Charest N, Benfenati E, Cronin MTD (2024) Molecular similarity in chemical informatics and predictive toxicity modeling: from quantitative read-across (q-RA) to quantitative read-across structure–activity relationship (q-RASAR) with the application of machine learning. Crit Rev Toxicol 54:659–684. https://doi.org/10.1080/10408444.2024.2386260
22. Roy K, Banerjee A (2024) q-RASAR: a path to predictive cheminformatics. Springer, New York. https://doi.org/10.1007/978-3-031-52057-0
23. Varsou D-D, Banerjee A, Roy J, Roy K, Savvas G, Sarimveis H, Wyrzykowska E, Balicki M, Puzyn T, Melagraki G, Lynch I, Afantitis A (2024) The round-robin approach applied to nanoinformatics: consensus prediction of nanomaterials zeta potential. Beilstein J Nanotechnol 15:1536–1553. https://doi.org/10.3762/bjnano.15.121
24. Gramatica P, Giani E, Papa E (2007) Statistical external validation and consensus modeling: a QSPR case study for Koc prediction. J Mol Graph Model 25:755–766. https://doi.org/10.1016/j.jmgm.2006.06.005
25. Katritzky AR, Sild S, Lobanov V, Karelson M (1998) Quantitative structure−property relationship (QSPR) correlation of glass transition temperatures of high molecular weight polymers. J Chem Inf Comput Sci 38:300–304. https://doi.org/10.1021/ci9700687
26. Khan PM, Roy K (2018) QSPR modelling for prediction of glass transition temperature of diverse polymers. SAR QSAR Environ Res 29:935–956. https://doi.org/10.1080/1062936X.2018.1536078
27. Rojas C, Duchowicz PR, Tripaldi P, Diez RP (2015) QSPR analysis for the retention index of flavors and fragrances on a OV-101 column. Chemom Intell Lab Syst 140:126–132. https://doi.org/10.1016/j.chemolab.2014.09.020
28. Gramatica P, Cassani S, Roy PP, Kovarich S, Yap CW, Papa E (2012) QSAR modeling is not "push a button and find a correlation": a case study of toxicity of (benzo-)triazoles on algae. Mol Inform 31:817–835. https://doi.org/10.1002/minf.201200075
29. Tropsha A (2010) Best practices for QSAR model development, validation, and exploitation. Mol Inform 29:476–488. https://doi.org/10.1002/minf.201000061
30. Mansouri K, Moreira-Filho JT, Lowe CN, Charest N, Martin T, Tkachenko V, Judson R, Conway M, Kleinstreuer NC, Williams AJ (2024) Free and open-source QSAR-ready workflow for automated standardization of chemical structures in support of QSAR modeling. J Cheminform 16:19. https://doi.org/10.1186/s13321-024-00814-3
31. Banerjee A, Roy K (2025) The multiclass ARKA framework for developing improved q-RASAR models for environmental toxicity endpoints. Environ Sci Processes Impacts 27:1229–1243. https://doi.org/10.1039/D5EM00068H
32. Crespo Márquez A (2022) The curse of dimensionality. In: Digital maintenance management. Springer series in reliability engineering. Springer, Cham. https://doi.org/10.1007/978-3-030-97660-6_7

33. Toth G, Bodai Z, Heberger K (2013) Estimation of influential points in any data set from coefficient of determination and its leave-one-out cross-validated counterpart. J Comput Aided Mol Des 27:837–844. https://doi.org/10.1007/s10822-013-9680-4

34. Van Drie JH (2003) Pharmacophore discovery—lessons learned. Curr Pharm Des 9:1649–1664. https://doi.org/10.2174/1381612033454568

35. Kolmar SS, Grulke CM (2021) The effect of noise on the predictive limit of QSAR models. J Cheminform 13:92. https://doi.org/10.1186/s13321-021-00571-7

36. Preetha UP, Suresh M, Tolasa FT, Bonyah E (2024) QSPR/QSAR study of antiviral drugs modeled as multigraphs by using TI's and MLR method to treat COVID-19 disease. Sci Rep 14:13150. https://doi.org/10.1038/s41598-024-63007-w

37. Abdullayev R, Khan K, Jillella GK, Nair VG, Amin SA, Roy J, Bousily M, Gajewicz-Skretna A (2025) Metal bioaccumulation prediction via QSPR-q-RASPR synergy and cross-species risk analysis. Green Chem 27:9679–9695. https://doi.org/10.1039/D5GC00904A

38. Banerjee A, Roy K (2024) ARKA: a framework of dimensionality reduction for machine-learning classification modeling, risk assessment, and data gap-filling of sparse environmental toxicity data. Environ Sci Processes Impacts 26:991–1007. https://doi.org/10.1039/D4EM00173G

39. Gajewicz A (2017) What if the number of nanotoxicity data is too small for developing predictive Nano-QSAR models? An alternative read-across based approach for filling data gaps. Nanoscale 9:8435–8448. https://doi.org/10.1039/C7NR02211E

40. Topliss JG, Costello RJ (1972) Chance correlations in structure-activity studies using multiple regression analysis. J Med Chem 15:1066–1068. https://doi.org/10.1021/jm00280a017

41. Tetko IV, van Deursen R, Godin G (2024) Be aware of overfitting by hyperparameter optimization! J Cheminform 16:139. https://doi.org/10.1186/s13321-024-00934-w

42. Fushiki T (2009) Estimation of prediction error by using K-fold cross-validation. Stat Comput 21:137–146. https://doi.org/10.1007/s11222-009-9153-8

43. Doweyko AM (2008) QSAR: dead or alive? J Comput Aided Mol Des 22:81–89. https://doi.org/10.1007/s10822-007-9162-7

44. Szeghalmy S, Fazekas A (2023) A comparative study of the use of stratified cross-validation and distribution-balanced stratified cross-validation in imbalanced learning. Sensors 23:2333. https://doi.org/10.3390/s23042333

45. Devilliers J, Doucet JP, Doucet-Panaye A, Decourtye A, Aupinel P (2012) Linear and non-linear QSAR modelling of juvenile hormone esterase inhibitors. SAR QSAR Environ Res 23:357–369. https://doi.org/10.1080/1062936X.2012.664562

46. Golbraikh A, Muratov E, Fourches D, Tropsha A (2014) Data set modelability by QSAR. J Chem Inf Model 54:1–4. https://doi.org/10.1021/ci400572x

47. Golbraikh A, Fourches D, Sedykh A, Muratov E, Liepina I, Tropsha A (2014) Modelability criteria: statistical characteristics estimating feasibility to build predictive QSAR models for a dataset. In: Leszczynski J, Shukla M (eds) Practical aspects of computational chemistry III. Springer, Boston, MA. https://doi.org/10.1007/978-1-4899-7445-7_7

48. Ruiz IL, Gomez-Nieto MA (2018) Study of data set modelability: modelability, rivality, and weighted modelability indexes. J Chem Inf Model 58:1798–1814. https://doi.org/10.1021/acs.jcim.8b00188

49. Banerjee A, Roy K (2022) First report of q-RASAR modeling toward an approach of easy interpretability and efficient transferability. Mol Divers 26:2847–2862. https://doi.org/10.1007/s11030-022-10478-6

50. Banerjee A, Roy K (2023) Prediction-inspired intelligent training for the development of classification read-across structure–activity relationship (c-RASAR) models for organic skin sensitizers: assessment of classification error rate from novel similarity coefficients. Chem Res Toxicol 36:1518–1531. https://doi.org/10.1021/acs.chemrestox.3c00155

Chapter 2
Outliers and the Applicability Domain of QSAR Models

Abstract Among the important determinants of the quality of the final models are outlier compounds and the applicability domain of the developed models. Outlier compounds may generally be of two types: structural outliers (sharing chemical characteristics uncommon to most of the remaining compounds) and prediction outliers (showing high residuals, unlike most of the remaining compounds). Identifying outliers is very important, as they may indicate important physicochemical features not yet considered during model development and may be a starting point for exploring a different mechanism of action. Some of the prediction outliers may be, in reality, activity cliffs, which show a significant difference in the activity values with compounds that are pretty similar in structural characteristics. When applying a QSAR model to predict external compounds, it is crucial to verify that they fall within the model's applicability domain, as no model can be universally applicable to all chemical compounds. The relevance of a QSAR model mainly depends on the chemical domain of the training compounds from which the model has learnt the structure-activity relationships.

Keywords Outliers · Applicability domain · Reliability of predictions · Uncertainty · Validation

2.1 Introduction to Applicability Domain and Outliers

The performance of a good QSAR model is based on its efficiency in accurately predicting the target endpoint of unseen query chemicals. This is possible when the training set possesses sufficient structural heterogeneity, and the model efficiently learns the intricate patterns and relationships of structural and physicochemical features with the target endpoint. Reliable predictions are generated when the input query molecule lies inside the chemical space encoded by the model with its selected features. This chemical space can be termed the Applicability Domain (AD) of a QSAR model [1], which is expected to encode the required chemical information of most of the query chemicals. However, in many cases, we find that some query

K. Roy, A. Banerjee, *Activity Cliffs*, SpringerBriefs in Molecular Science, https://doi.org/10.1007/978-3-032-10081-8_2

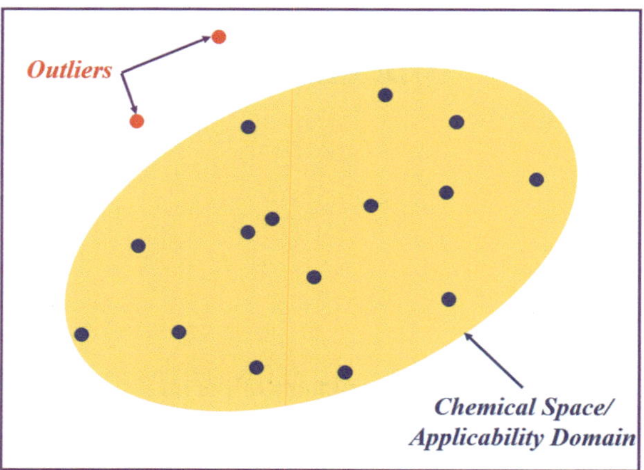

Fig. 2.1 A pictorial representation of the AD of a model, and the depiction of outliers

compounds lie outside this theoretical chemical space of the model, which are termed "Outliers" (Fig. 2.1). This can be explained from two different points of view—firstly, the query compound has a significantly different chemical structure and/or properties as compared to most other members affecting the target endpoint; secondly, the developed model has a limited domain of applicability, marking the target compound outside the domain. The second case arises when the QSAR modeler either used a limited number of data points to train a model, or the training set compounds have a high level of structural homogeneity, where the compounds are significantly similar to each other. If the query molecule lacks sufficient structural similarity to this homogenous training set, it will be marked as an outlier, even though this query molecule may have sufficient similarity to other molecules affecting the target endpoint, which apparently are not used to train the QSAR model. Therefore, the concepts of AD and the identification of outliers are very relative, and depend on the training set chemical and feature spaces, which is why they warrant the consideration of structural heterogeneity in the training set chemicals. The aim of a QSAR modeler is to develop a robust and highly predictive model, using a minimal number of descriptors but maximizing the coverage of a large chemical space. This not only enhances the statistical reliability of the developed model, but also aids in enhanced generalization and reliability towards unseen compound predictions.

In 2016, Hanser et al. [2] proposed the concept of the decision domain, which considers a three-step approach, namely the applicability domain, reliability domain, and decidability domain. These steps are essential and they enable the possibility to make decisions based on valid, reliable, and non-equivocal predictions. The first step is to check the validity of the developed model, in terms of its applicability domain, for the prediction of an unknown compound. If the model is found to be inappropriate for the intended prediction, then the query compound can be treated as the one, which is beyond the AD of the model (an outlier). If it is otherwise, i.e., the query compound is compatible with the model, then the query compound lies inside the AD. Once it is ensured that the query compound is within the AD, a question arises regarding the prediction reliability. The purpose of assessing reliability is to ensure and notify the user regarding the quality, quantity, and relevance of the information presented by the model. After ensuring that the input information is at the same level as the information presented by the model, it can be concluded that the prediction reliability is good. Decidability, on the other hand, is an estimation of the errors in an individual prediction. This is essential as it measures the degree to which one can trust the conclusion derived from a prediction. While reliability ensures trust in the prediction itself, decidability ensures trust in the conclusions derived from the prediction [3].

2.2   Outliers and Their Detection

Outliers are an important measure to judge the quality and coverage of the developed QSAR model. As stated previously, a QSAR modeler aims to cover a large chemical space, which makes unseen compounds less prone to being labelled outside the AD of the model. However, another aim of a QSAR modeler is to develop predictive models using a minimal number of descriptors. In reality, there is always a trade-off between these two concepts as the AD coverage requires the consideration of a greater number of descriptors, while statistical reliability warrants the use of a lower number of descriptors. However, indiscriminately considering a large feature pool introduces noise in the training set due to the "Curse of Dimensionality" [4] (as mentioned in Chap. 1), where overfitted models are developed. In such cases, the AD coverage of the model is significantly misled with respect to the target endpoint and is driven by the noise created by the irrelevant descriptors. Such models tend to have a poor predictivity on test/unseen data, and the poorly predicted compounds are labelled as "prediction outliers", apart from the fact that they are "structural outliers", with respect to the fuzzy AD. In other cases, when the modeler selects too few descriptors, the AD of the model becomes poorly defined. Unseen compounds, in such cases, may become structural outliers, although possessing the relevant structural features, as per literature knowledge, for a particular endpoint. This arises when the training set representatives do not have those particular structural features, which results in the definition of AD without considering those features. Therefore, the job of a modeler is to optimally balance these two aspects, thus ensuring a larger AD coverage without the unnecessary use of additional descriptors.

Structural outliers are detected using an unsupervised approach. In many case, the developed model does not understand the structure of an input query compound, as the requisite features present in the query compound have not been used for model development. This is particularly evident when the training dataset lacks sufficient structural heterogeneity, resulting in structurally homologous training set members. Figure 2.2 demonstrates how a query compound can be labelled as an outlier with respect to a homogeneous training set, while the same

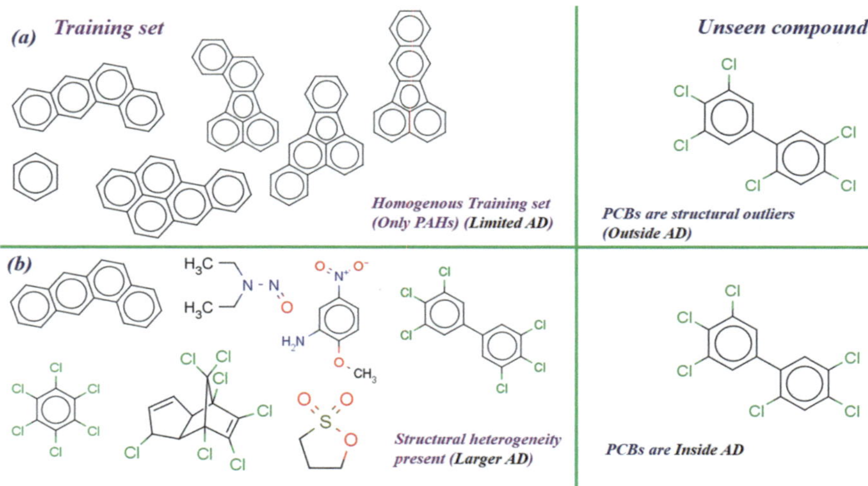

Fig. 2.2 Representation of (**a**) Homogeneous training set with a limited AD labels the query compound as an outlier, and (**b**) Heterogeneous training set with a larger AD labels the query compound as Inside AD

compound can be within the AD if the training set is structurally diverse. In Fig. 2.2a, the training set is homogenous, consisting of only Poly-Aromatic Hydrocarbons (PAHs). If the model developed using this particular training set tries to predict the endpoint (for example, Inhalation Slope Factor as a measure of carcinogenic potency) of a Poly-Chlorinated Biphenyl (PCB) compound, it will be labelled as a compound that is beyond the AD of this QSAR model. However, if the same endpoint is modeled, but with a training set of sufficient structural diversity/heterogeneity, it is most likely that the same PCB compound would be inside the AD of this model.

Prediction outliers, on the other hand, refers to the case where the predicted residuals for a particular query compound is quite high. In such cases with a large difference in the observed and predicted endpoint values, questions arise on the reliability of the prediction. This phenomenon may not be related to the difference in structural features of the query compound and the training set representatives, as prediction outliers are not always structural outliers. As explained by Maggiora [5], the identification of outliers may not always be based on statistical consideration, but rather the positioning of the data point in the structure-activity landscape needs to be explored. Therefore, valid data points that appear in the cliff region (activity cliffs) may also be outliers. As per the Best Practices of QSAR model development, the modeler should remove outliers prior to the development of a QSAR model, which enhances the latter's reliability [6]. In our opinion, the "outliers" to be removed during model development should include structural outliers and activity cliffs.

2.3   Different Measures of Applicability Domain

Assessment of AD for a QSAR model is an essential aspect. A particular QSAR model is not trained to predict literally every compound in the universe; instead, the focus should be on its relevance in terms of the defined chemical space. Among the various approaches that estimate the AD of a QSAR model, we will discuss the common ones in this section.

2.3.1   Bounding Box (Range-Based Approach)

This is one of the simplest approaches for the estimation of AD of a QSAR model. This approach considers the ranges in descriptors and response values, in terms of their individual maximum and minimum values in the training set, to define a "box". If a particular query molecule has descriptor and/or response values "beyond" the training set ranges, such compounds can be tagged outliers [7]. The drawback of this approach is that when the training set data are not uniformly distributed, this approach fails to explore the empty regions in the interpolated space.

2.3.2   PCA Bounding Box (Range-Based Approach)

This is a similar approach to the conventional bounding box. The descriptor space undergoes unsupervised dimensionality reduction using Principal Component Analysis (PCA). The lowest and the highest values of each PC are used to define the range. The data points are plotted in a PC1 vs. PC2 plot, and the AD is defined by a circle. Query compounds lying beyond this circle can be classified as outliers.

2.3.3   Convex Hull (Geometrical Method)

Convex hull is based on a computational geometry-based approach. It uses an n-dimensional dataset and estimates its coverage using the convex hull calculations [8]. This approach defines the boundary of the dataset based on the degree of data distribution. Convex hull is more suited to data having two or three dimensions; however, the complexity increases exponentially as the input data dimension increases.

2.3.4   *Leverage Approach (Distance-Based Method)*

The leverage approach is one of the most widely used approaches for AD determination [9]. It uses the n-dimensional descriptor matrix to compute the HAT matrix, as shown in Eq. 2.1.

$$HAT = \left(X^T \left(X^T X \right)^{-1} X \right) \tag{2.1}$$

In Eq. 2.1, X represents the n-dimensional descriptor matrix, and X^T represents the transposed descriptor matrix. $(X^T X)^{-1}$ is called the covariance matrix. For the calculation of the training and test set leverages, X and X^T values change accordingly, as the input feature matrices for the training and test sets are different. However, the covariance matrix value should always and exclusively be computed based on the training set descriptor matrix. The HAT matrix typically results in the generation of a square matrix of dimension $n \times n$ (where n is the number of data points), whose diagonal values, from left to right, represent the leverages. A threshold leverage value (h^*) is also calculated using the formula (Eq. 2.2), where p is the number of descriptors.

$$h^* = \frac{3(p+1)}{n} \tag{2.2}$$

Structural outliers are those compounds that have leverage values greater than the threshold leverage, while prediction outliers are those compounds that have standardized predicted residuals values beyond ± 3 log units. For the training set compounds, this standardized predicted residual arises from the difference between the standardized experimental response values and the cross-validated standardized predicted response values. A Williams Plot visually depicts the structural and prediction outliers identified using the leverage approach (Fig. 2.3).

2.3.5   *Euclidean Distance-Based Approach*

Euclidean distance (ED) is one of the most common measures to evaluate the distance between compounds in a particular chemical space. This is a useful indicator to define the AD of a model [10]. The Euclidean distance between two different data points X and Y can be represented as Eq. 2.3.

$$d(X,Y) = \sqrt{\sum_{i=0}^{n} (X_i - Y_i)^2} \tag{2.3}$$

Fig. 2.3 A Williams Plot (Standardised residuals vs. Leverages) representing the Inside AD and Outlier compounds

In Eq. 2.3, d(X,Y) is the Euclidean distance between data points X and Y, X_i represents the ith descriptor of compound X, and Y_i represents the ith descriptor of compound Y.

The mean ED of a target compound with respect to all the other compounds in the dataset can be represented as in Eq. 2.4.

$$\overline{d_{(X,Y)}} = \frac{\sum_{n-1}^{j=1} d(X,Y)_j}{n-1} \tag{2.4}$$

Equation 2.4 represents the mean ED of a particular query compound with respect to n number of training compounds. This mean distance is then normalized from 0 to 1 for each training compound. The maximum and minimum distances are then used to normalize the mean ED of test/unseen compounds. If the normalized mean ED values for the unseen compounds are beyond 1, then these compounds can be classified as outside AD.

2.3.6   Mahalanobis Distance-Based Approach

In 1936, Prasanta Chandra Mahalanobis introduced the Mahalanobis distance, which measures the distance between a data point X and a probability distribution P [11]. The basic theory of the Mahalanobis distance-based AD approach is that it computes the distance of an observation to the mean values of the descriptors, not considering the prediction values. The observations that are higher than the remaining ones can be labelled Outside AD. Mathematically, Mahalanobis distance can be represented as in Eq. 2.5, where x_i and x_j are two different data points.

$$d_{Mahalanobis}\left(x_i, x_j\right) = \sqrt{\left(x_i - x_j\right)^T \Sigma^{-1}\left(x_i - x_j\right)} \tag{2.5}$$

2.3.7 Hotelling T^2 Test

It is a multivariate form of the Student's t-test and related to the Mahalanobis distance and leverage-based approaches. It presumes that the data is normally distributed and identifies the significance of mean differences of two or more variables between two groups. The covariance matrix is used to correct for the collinear descriptors and measures the distance of a query compound from the center of a set of observations. The t-value decides whether the query compound is inside or outside AD.

2.3.8   k-Nearest Neighbors-Based Approach

This is a basic approach where the similarities of a query compound to its close k-neighbors determines the reliability in predictions [12]. If the nearest neighbors appear to be very close to the query compound in similarity space, it can be inferred that the query compound has high reliability. However, if the nearest neighbors appear to be very far, and the query compound appears to be a singleton in the similarity space, it can be inferred that the query compound has low reliability and low structural similarity with the training set congeners.

2.3.9   DModX, a.k.a., Distance to the Model in X-Space

Distance to the Model in X-space (DModX) is a typical measure to estimate the AD in the case of a Partial Least Squares (PLS) model [13]. It indicates how far a particular observation is situated from the model plane, in terms of residual standard deviation between the latent variables and the data [14]. If the DModX value of a compound is greater than the critical calculated value (DCrit), it can be termed as an outlier.

2.3.10   Tanimoto Similarity-Based Approach

Tanimoto, a.k.a. Jaccard similarity index, is a measure that takes into account the number of common features between a pair of molecules. It is a useful and efficient similarity measure particularly when molecular fingerprints define the feature space. If the Tanimoto similarity of a query compound is low, as compared to the training set representatives, such query compounds can be labelled outliers [15].

2.3.11   Standardization Technique

This is an approach proposed by Roy et al. [16]. The training set descriptors are assumed to be normally distributed with an insignificant skewness, and the majority of the data points should lie in the region of mean ±3× standard deviation. Any data point lying beyond this boundary can be termed outliers, since such compounds are structurally very different from majority of the members in the dataset. The absolute standardized descriptor values (S) of all descriptors for a particular compound are computed. If the maximum absolute standardized descriptor value for a compound is ≤3, then the compound is inside AD. On the other hand, if the absolute minimum standardized descriptor value for the compound is >3, then the compound can be classified as an outlier. However, in cases where the maximum absolute standard deviation value is greater than 3, but the minimum absolute standard deviation value is ≤3, then S_{new} for the compound k can be calculated using Eq. 2.6.

$$S_{new(k)} = \overline{S_k} + 1.28 \times \sigma_{S_k} \tag{2.6}$$

In Eq. 2.6, $\overline{S_k}$ represents the mean of the absolute standardized descriptor values of all descriptors for the compound k, and σ_{S_k} represents the corresponding standard deviation. If the value of $S_{new(k)}$ is ≤3, then the compound is within the AD.

2.3.12   The DTC Plot to Determine Prediction Confidence Outliers

Banerjee and Roy proposed this approach in 2023 [17]. It considers the similarity and activity data for the first two closest neighbors for each query compound. In this plot, the Y-axis represents the absolute difference in activity values of the first two closest source compounds, the X-axis represents the compound IDs that are arranged in descending order (from left to right) of their highest similarity value to their closest source neighbors, and the bubble diameter represents the absolute difference in the similarity values of the first two closest source compounds. Ideally, for a particular query compound, a larger absolute difference in similarity values of the first two closest source congeners should also have a large absolute difference in their observed activity values. Therefore, data points with a larger bubble diameter, located higher on the Y-axis, and towards the left of the plot (where the similarity value to the nearest neighbor is very high), and vice versa, signify confident data points. However, data points with a considerable level of absolute similarity difference but a large absolute difference in the observed activity values of the first two

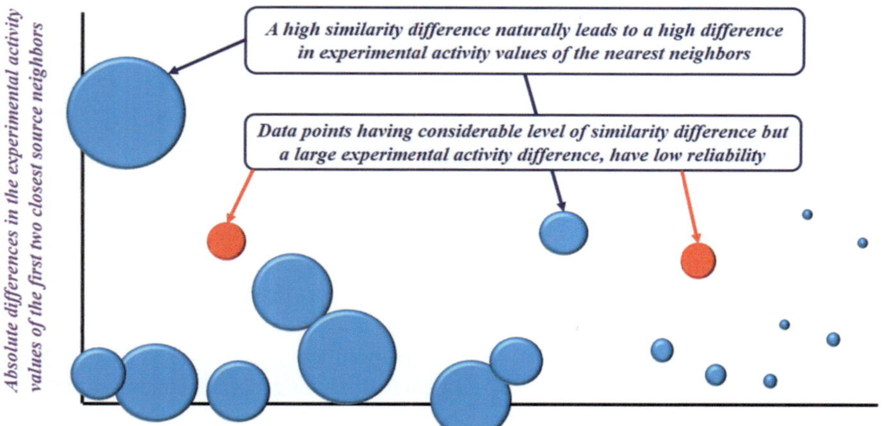

Fig. 2.4 The DTC Plot to demonstrate the q-RASAR model Applicability Domain

nearest neighbors can have reduced reliability in terms of similarity. Although this has been explained as a part of a q-RASAR modeling perspective in the original paper, the concept is universally applicable from a QSAR modeler's point of view. A representative figure of the DTC plot has been provided in Fig. 2.4.

2.3.13 AD Analysis for Machine Learning Models

2.3.13.1 SHAP Leverage Approach

SHAP, a.k.a. SHapley Additive exPlanations, is an algorithm based on game theory, which is used to identify the feature contributions in a Machine Learning model [18]. Recently, Makarov et al. [19] employed the concepts of the leverage approach to compute the AD using SHAP values. They stated that the computation of conventional leverages for non-linear models may be over-optimistic, resulting in incorrect AD assessment. For a particular query compound, the authors computed the SHAP values of each feature and defined leverage (h) as the summation of the absolute SHAP values for a particular data point i (Eq. 2.7).

$$h_i = \sum_{j=1}^{d} \left| SHAP_{ij} \right| \tag{2.7}$$

In Eq. 2.7, $SHAP_{ij}$ denotes the contribution of the jth feature/descriptor to the ith data point, while d represents the total number of descriptors. The threshold leverage (h^*) was set at the 95th percentile of the leverage distribution.

2.3.13.2   Optimization of AD Methods for Machine Learning Models

Kaneko [20] presented an approach for the optimization of AD methods. The author employed three different AD methods across various datasets. At first, the authors generated predictions using a double cross-validation approach. The computed Root Mean Squared Error (RMSE) was plotted against coverage, and the Area Under the Coverage and RMSE curve (AURC) was developed. The best AD method and hyperparameter were selected, which minimized the AUCR.

2.3.13.3   Sum of the Distance-Weighted Contributions (SDC)

This is an approach proposed by Liu and Wallqvist [21]. The authors stated that defining AD based on ensemble variance (i.e., variances in predictions from an ensemble QSAR model) may fail to capture large prediction errors in melting point data, for which they proposed the SDC metric (Eq. 2.8).

$$SDC = \sum_{i=1}^{n} e^{-3TD_i/(1-TD_i)} \tag{2.8}$$

In the above equation, TD refers to the Tanimoto distance from a query molecule (n) to the ith training compound. This Tanimoto distance was calculated using the ECFP_4 circular fingerprint. The authors made a comparative analysis in which they used a set of compounds having melting points $>0\ °C$ to train a Random Forest model to predict a test set having melting points $<0\ °C$. As expected, the test set compounds were outside the AD with respect to the training set, and most of these compounds have SDC values of zero. This was plotted in a prediction error vs. SDC plot. Additionally, when the authors expanded the training set with additional data points, the SDC values seemed to increase for many data points, resulting in the generation of a lower number of out-of-AD compounds.

Additionally, estimation of class probabilities for the query compounds can also act as an indicator of the AD assessment [22].

2.4   Uncertainty in QSAR Predictions

The uncertainty of QSAR predictions is significant to define, as this is related to the reliability of the predictions for the query compounds. The uncertainty can be expressed qualitatively, in terms of prediction reliability (which is mainly done by defining the applicability domain), and quantitatively in terms of a measure of errors in predictions [23]. Extrapolation of the model domain for making predictions influences both the quantitative and qualitative uncertainties. The model domain primarily relates to the chemical structural space defined by molecular descriptors and fingerprints. Still, it may also encompass the domains specified by mode of action and metabolic aspects. On the other hand, QSAR predictions with uncertainty can be obtained by deriving the probabilistic measures of predictive performance, for example, using Bayesian statistics (especially for small data sets) or by using an appropriate sampling/resampling technique (suitable for large data sets). A non-probabilistic measure of predictive uncertainty is the standard deviation in ensemble predictions or ensemble metrics for predictions.

Resampling and error modelling-based uncertainty estimation techniques consider the observations and predictions as distributions. The uncertainty estimates obtained from each method are applied to convert predictions to Gaussian distributions. A likelihood-based measure is used to assess the performance of the alternative techniques. Kullback–Leibler (KL) divergence, which measures the distance between two probability distributions, can be used to assess the quality of a set of predictive distribution outputs from a model [24].

Conformal Prediction (CP) is a nonparametric method that can be used for the estimation of confidence of predictions in a non-specific manner to the modelling algorithm [25]. The method produces a prediction interval corresponding to a confidence threshold set by the user.

The prediction error for a new molecule depends on, in addition to the selected features and modeling algorithms, its distance to the training molecules, which may serve as the basis for developing a high-performance reliability metric. Liu et al. [26] introduced a Tanimoto distance-based metric, known as the sum of distance-weighted contributions (SDC), which correlated well with the prediction errors for different data sets.

Roy et al. [27] categorized the reliability of predictions for query compounds into Good, Moderate, and Poor, based on three different criteria: cross-validated mean absolute error of the ten closest training compounds, the applicability domain based on the model features, and the proximity of predictions to the training set response mean. They also reported a tool, the "Prediction reliability Indicator", which can be used to determine the confidence in predictions from MLR/PLS models; however, the same concept can easily be extended for predictions obtained from models using other algorithms.

For a large and heterogeneous data set, the quality of predictions may also vary in the different regions of chemical space, which may be a fascinating field of study. While it is mandatory to report whether the query compound falls within the applicability domain of a predictive model in the regulatory context [23], it seems also essential to indicate the quantitative uncertainty related to the prediction accuracy or errors.

References

1. Gadaleta D, Mangiatordi GF, Catto M, Carotti A, Nicolotti O (2016) Applicability domain for QSAR models: where theory meets reality. Int J Quant Struct-Prop Relat 1:45–63. https://doi.org/10.4018/IJQSPR.2016010102

2. Hanser T, Barber C, Marchaland JF, Werner S (2016) Applicability domain: towards a more formal definition. SAR QSAR Environ Res 27:865–881. https://doi.org/10.108 0/1062936X.2016.1250229

3. Kar S, Roy K, Leszczynski J (2018) Applicability domain: a step toward confident predictions and decidability for QSAR modeling. In: Nicolotti O (ed) Computational toxicology. Methods in molecular biology, vol 1800. Humana Press, New York, NY. https://doi.org/10.1007/978-1-4939-7899-1_6

4. Crespo Márquez A (2022) The curse of dimensionality. In: Digital maintenance management. Springer series in reliability engineering. Springer, Cham. https://doi.org/10.1007/978-3-030-97660-6_7

5. Maggiora GM (2006) On outliers and activity cliffs—why QSAR often disappoints. J Chem Inf Model 46:1535. https://doi.org/10.1021/ci060117s

6. Tropsha A (2010) Best practices for QSAR model development, validation, and exploitation. Mol Inform 29:476–488. https://doi.org/10.1002/minf.201000061

7. Jaworska J, Nikolova-Jeliazkova N, Aldenberg T (2005) QSAR applicability domain estimation by projection of the training set in descriptor space: a review. Altern Lab Anim 33:445–459. https://doi.org/10.1177/026119290503300508

8. Preparata FP, Shamos MI (1991) Computational geometry: an introduction. Springer, New York. https://doi.org/10.1007/978-1-4612-1098-6

9. Gramatica P (2007) Principles of QSAR models validation: internal and external. QSAR Comb Sci 26:694–701. https://doi.org/10.1002/qsar.200610151

10. Minovski N, Zuperl S, Drgan V, Novic M (2013) Assessment of applicability domain for multivariate counter-propagation artificial neural network predictive models by minimum Euclidean distance space analysis: a case study. Anal Chim Acta 759:28–42. https://doi.org/10.1016/j.aca.2012.11.002

11. Mahalanobis PC (1936) On the generalised distance in statistics. Proc Natl Inst Sci India 2:49–55. https://doi.org/10.1007/s13171-019-00164-5

12. Sheridan R, Feuston RP, Maiorov VN, Kearsley S (2004) Similarity to molecules in the training set is a good discriminator for prediction accuracy in QSAR. J Chem Inf Comput Sci 44:1912–1928. https://doi.org/10.1021/ci049782w

13. Wold S, Sjostrom M, Eriksson L (2001) PLS-regression: a basic tool of chemometrics. Chemom Intell Lab Syst 58:109–130. https://doi.org/10.1016/S0169-7439(01)00155-1

14. Massei A, Falco N, Fissore D (2025) NIR-based real-time monitoring of freeze-drying processes: application to fault and endpoint detection. Processes 13:452. https://doi.org/10.3390/pr13020452

15. Tetko IV, Sushko I, Pandey AK, Zhu H, Tropsha A, Papa E, Oberg T, Todeschini R, Fourches D, Varnek A (2008) Critical assessment of QSAR models of environmental toxicity against Tetrahymena pyriformis: focusing on applicability domain and overfitting by variable selection. J Chem Inf Comput Sci 48:1733–1746. https://doi.org/10.1021/ci800151m

16. Roy K, Kar S, Ambure P (2015) On a simple approach for determining applicability domain of QSAR models. Chemom Intell Lab Syst 145:22–29. https://doi.org/10.1016/j.chemolab.2015.04.013

17. Banerjee A, Roy K (2023) Machine-learning-based similarity meets traditional QSAR: "q-RASAR" for the enhancement of the external predictivity and detection of prediction confidence outliers in an hERG toxicity dataset. Chemom Intell Lab Syst 237:104829. https://doi.org/10.1016/j.chemolab.2023.104829

18. Rodriguez-Perez R, Bajorath J (2019) Interpretation of compound activity predictions from complex machine learning models using local approximations and Shapley values. J Med Chem 63:8761–8777. https://doi.org/10.1021/acs.jmedchem.9b01101

19. Makarov DM, Kalikin NN, Budkov YA, Gurikov P, Kruchinin SE, Jouyban A, Kiselev MG (2025) Improved solubility predictions in scCO2 using thermodynamics-informed machine learning models. J Chem Inf Model 65:4043–4056. https://doi.org/10.1021/acs.jcim.5c00432

20. Kaneko H (2024) Evaluation and optimization methods for applicability domain methods and their hyperparameters, considering the prediction performance of machine learning models. ACS Omega 9:11453–11458. https://doi.org/10.1021/acsomega.3c08036

21. Liu R, Wallqvist A (2019) Molecular similarity-based domain applicability metric efficiently identifies out-of-domain compounds. J Chem Inf Model 59:181–189. https://doi.org/10.1021/acs.jcim.8b00597
22. Klingspohn W, Mathea M, ter Laak A, Heinrich N, Bauman K (2017) Efficiency of different measures for defining the applicability domain of classification models. J Cheminform 9:44. https://doi.org/10.1186/s13321-017-0230-2
23. Sahlin U (2013) Uncertainty in QSAR predictions. Altern Lab Anim 41:111–125. https://doi.org/10.1177/026119291304100111
24. Wood DJ, Carlsson L, Eklund M, Norinder U, Stalring J (2013) QSAR with experimental and predictive distributions: an information theoretic approach for assessing model quality. J Comput Aided Mol Des 27:203–219. https://doi.org/10.1007/s10822-013-9639-5
25. Papadopoulos H, Vovk V, Gammerman A (2011) Regression conformal prediction with nearest neighbours. J Artif Intell Res 40:815–840. https://doi.org/10.1613/jair.3198
26. Liu R, Glover KP, Feasel MG, Wallqvist A (2018) General approach to estimate error bars for quantitative structure–activity relationship predictions of molecular activity. J Chem Inf Model 58:1561–1575. https://doi.org/10.1021/acs.jcim.8b00114
27. Roy K, Ambure P, Kar S (2018) How precise are our quantitative structure–activity relationship derived predictions for new query chemicals? ACS Omega 3:11392–11406. https://doi.org/10.1021/acsomega.8b01647

Chapter 3
Cliffs in Biological Activity Landscape

Abstract The structure-activity landscape of chemical compounds may not always be smooth, though this is a fundamental assumption of QSAR modeling. There may be several cliffs in the landscape, where some chemical compounds may show different activity (of the order 100-fold or more) than other close congeners in the series. In such cases, the predictions for such compounds derived from QSAR would naturally be disappointing, as the similarity principle is not followed here. Thus, identifying activity cliffs is crucial for developing statistically acceptable QSAR models. The definition of similarity for identifying activity cliffs may be based on chemical fingerprints or descriptors (classical activity cliffs), substructures (chirality cliffs, matched molecular pair cliffs), three-dimensional structure-based (3D-cliffs), or the target-set dependent potency difference. Some prediction outliers, even within the applicability domain of QSAR models, may arise due to the activity cliff (AC) behavior. In addition to compound pairs, activity cliffs can also be visualized in coordinated networks that form AC clusters. Despite using high-quality data, the data set's modelability may be significantly compromised in the presence of ACs, among other factors. Different methods for identifying activity cliffs have been proposed, such as the structure-activity landscape index (SALI), structure-activity relationship index (SARI), and machine-learning-based methods.

Keywords Activity cliffs · Modelability · Predictions · QSAR · SALI · SARI · ARKA

3.1 Introduction to the Concept of Activity Cliffs

Activity cliffs (ACs) can be defined as pairs or groups of compounds that share a high level of structural similarity but possess opposite or large differences in activity/potency against a particular target [1, 2]. Such compounds tend to oppose the generalized notion of the similarity principle, which states that similar compounds should show similar biological activities [3]. In the context of QSAR modeling,

K. Roy, A. Banerjee, *Activity Cliffs*, SpringerBriefs in Molecular Science, https://doi.org/10.1007/978-3-032-10081-8_3

where SAR stands for "Structure-Activity Relationships", it becomes imperative to identify and remove such activity cliffs before model development, as these compounds tend to mislead the efficient training of the model. To the best of our knowledge, there have been a limited number of studies identifying activity cliffs and exploring their nature using suitable methods. However, the work of Jurgen Bajorath on activity cliffs suggests that there is much to explore, and this may justify some of the incorrect model-derived predictions [4–6].

The concept of activity cliffs is relative, as the structural similarity of molecules depends on the input feature matrix. It may be possible that the considered feature matrix lacks key descriptors/fingerprints essential to distinguish activity cliff pairs. Additionally, the identification of ACs depends on the level of information the feature matrix encodes (i.e., whole molecule-based similarity or substructure-based similarity) [1, 5]. The following section details the various types and generations of activity cliffs based on the information content of the feature matrix and target specificity.

3.2 The Nature and Types of Activity Cliffs

The appropriate definition of similarity among molecules greatly influences their activity cliff nature. As a standard practice in QSAR modeling, we consider the whole molecular similarity for model development and prediction purposes, based on the similarity principle. This whole molecular similarity can be defined in various ways, depending on the types and the number of selected features. As a general notion, as the number of features increases, the similarity between compound pairs tends to decrease. However, it should be noted that the similarity should ideally be defined based on features that have a relation to the target endpoint. In many cases, Tanimoto similarity is computed based on binary fingerprint representations and is used to identify the common fingerprint bits between AC pairs. A greater number of overlaps between fingerprint bits typically represents that the compounds are similar to each other. However, the definition of a "similarity threshold" varies, as different fingerprint representations yield varying similarity levels. Similarly, this is also the case when other similarity measures, such as cosine similarity or Euclidean distances, are used to estimate the similarity between molecules. Therefore, the similarity is "relative" and is entirely dependent on the feature space and the similarity measure employed. Additionally, there are interpretability issues associated with Tanimoto similarity, as it is restricted to the whole molecular similarity and fails to consider reaction information and substituent rules [1]. Alternatively, substructure-based similarity may be considered.

The substructure-based similarity is based on whether compounds share a common pre-defined substructure or not. The presence of different scaffolds and substituents can be used to define substructure-based similarity. As described by Bajorath's group [4, 7], this substructure-based similarity can be divided into three categories. Category 1 represents AC pairs having the same scaffold and R-groups. This includes topology cliffs, where the position of a particular functional group varies, and chirality cliffs that are affected by stereochemistry. It is evident that descriptors with lower dimensions, such as 1D fingerprints or 0-2D structural descriptors, are insufficient for distinguishing chirality cliffs, as they do not encode 3D stereochemistry. Therefore, effectively Quinine and Quinidine are the same compounds for a QSAR model developed using the lower dimensions of descriptors, which fail to take into account their wide variety of activity spectrum, where Quinine is antimalarial and Quinidine is antiarrhythmic. On the other hand, Category 2 represents AC pairs that have the same molecular scaffold, but differ in the R-groups. AC pairs may not only possess different R-groups attached to the same scaffold, but also the number of R-groups attached to the scaffold may vary. This category of ACs (R-group cliffs) occurs with higher propensity as compared to the other types of ACs [4]. Category 3 represents AC pairs having different scaffolds but the same R-groups. Such AC pairs typically have the same types and number of R-groups, but they are attached to different scaffolds and are therefore referred to as scaffold cliffs. The formation of matched molecular pairs (MMPs), i.e., a pair of compounds that only differ by a chemical modification at a particular site, is also a substructure-based similarity criterion, and such compounds are labelled MMP-cliffs.

Speaking more on the importance of encoding a higher level of information, especially from a drug discovery point of view, the 3D similarity is often considered for identifying 3D cliffs. When ligand molecules bind to the active site of target receptor/proteins, the 3D similarity can be calculated based on the positional differences between different ligand molecules. 3D-cliffs emerge when different ligands, although possessing similar binding modes, have large differences in their potencies.

Activity cliffs not only occur in pairs but also mainly as a group of similar structural analogs. These AC "clusters" are composed of a highly influential/potent compound, along with multiple weakly potent compounds that form a network [8]. Again, from a similarity point of view, the nearest neighbors are the ones that influence the target compound the most. As the distance between neighbors increases, their effect on the target compound decreases. Recently, Banerjee and Roy [9, 10] demonstrated how the opposite activity status of the nearest neighbors facilitated the identification of the target compound as an activity cliff, which explained why the model mispredicted these compounds.

The identification of ACs is often target-dependent, and it is essential to define a threshold for labelling compounds as ACs. Bajorath's group inferred that it is necessary to define a set-dependent threshold criterion for identifying target set-dependent ACs [11, 12]. Initially, the compounds forming analog pairs were identified, and their absolute differences in potency were calculated. The threshold was defined as the mean absolute potential difference + 2 times the standard deviation. Analog pairs having their mean absolute potency difference beyond the threshold were classified as target set-dependent ACs.

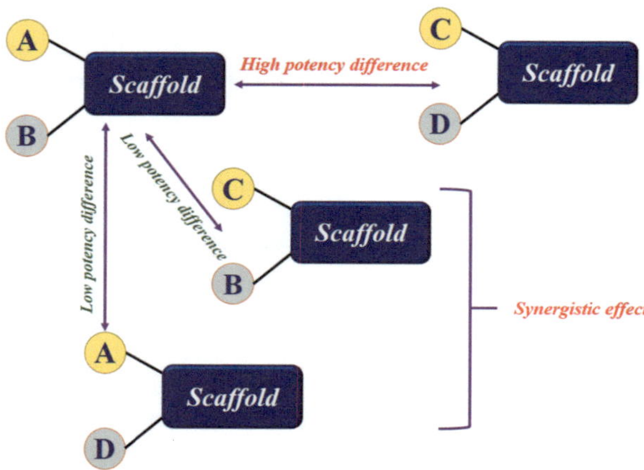

Fig. 3.1 Third-generation dual-site ACs

Activity cliffs can be classified into different generations. The first-generation activity cliffs are characterized by fingerprints and descriptors, similarity metrics like the Tanimoto similarity, and a defined potency threshold [13]. The second generation of activity cliffs is characterized by the consideration of a substructure-based similarity criterion, like the generation of MMPs, and using a target set-dependent difference in potency criteria among analog pairs [5, 12]. From a medicinal chemist's point of view, the use of MMPs often increases the ease of interpretability, while the target set-dependent potency difference criteria enhance the SAR information content. Third-generation activity cliffs are based on varying target set-dependent potency difference thresholds [14]. This involves modification/ substitution at single and multiple sites (dual-site ACs) of analog pairs. Considering a pair of dual-site ACs, the single-site analogs are identified. These analogs have typical single-site substitutions on the same molecular scaffold. The single-site substitutions act synergistically to generate ACs (Fig. 3.1).

3.3   The Methods of Detection of Activity Cliffs

Identification of ACs can be very challenging for cheminformatics researchers, as they do not obey the similarity principle. While the concept of the similarity principle acts as a pillar for the Structure-Activity Relationships (SAR) of compounds, it is therefore essential to identify cliffs to generate a more explainable version of the SAR. In this section, we will discuss the various methodologies proposed by different research groups for the efficient identification of ACs, aimed towards enhancing the modelability of a dataset.

3.3.1   The SALI Index

The Structure-Activity Landscape Index (SALI), developed by Guha and Van Drie, is an index that helps in the identification of ACs [15]. This is one of the simplest representations, highlighting the case where analog pairs do not seem to obey the similarity principle. Mathematically, this can be represented as in Eq. 3.1.

$$SALI_{i,j} = \frac{|A_i - A_j|}{1 - sim(i,j)} \tag{3.1}$$

In Eq. 3.1, A_i and A_j represent the activity values of the ith and the jth molecules. Again, according to the similarity principle, a lower similarity among analog pairs attributes to a large difference in experimental response values, and vice versa. However, in cases where both the absolute activity difference and the similarity between analog pairs are high, such compounds can be classified as ACs. Therefore, a higher SALI value should be attributed to the activity cliff nature of molecules. The authors also state that the SALI value may become infinite in cases where the similarity value between two compounds becomes 1, as the denominator section of the SALI equation becomes 0. In such cases, the authors replaced such SALI values with the next largest SALI value. It should be noted that the similarity value between two non-identical compounds may become 1, as (i) the similarity is defined based on certain selected features, and (ii) in cases of stereoisomers. In the introductory paper of SALI [15], the authors computed the similarity using the Tanimoto coefficient computed on fingerprint pairs.

Additionally, the authors noted the importance of the visualization of SAR. For a set of M compounds, it is possible to generate an M × M matrix of SALI values. Sorting the SALI values from highest to lowest will show the compounds that are "most confident" activity cliffs, which have high similarity and the largest change in activities. However, the authors stated that this approach is a rather "local" view of SAR, which necessitates the development of a more "global" view. This global view was generated by representing the compounds as nodes, which are then connected. This connection only occurs when the SALI values between the two analog pairs are greater than the user-defined threshold/cutoff SALI value. As this cut-off is decreased, a greater number of connections can be seen connecting different nodes, and vice versa. Therefore, if the user wants to estimate only the most significant ACs, the cut-off value may be increased. According to the authors, the goal of setting a user-defined cut-off is to enable users to identify the increasingly significant ACs.

Generating a heat map can be another method to represent the global view of SAR. The authors [15] plotted a 2D heatmap, where the x and y axes represented the data points in increasing order of their activity. The color intensity was depicted according to the SALI values between paired data points (lower SALI values corresponded to darker blocks). Lighter blocks localized near the upper left corner of the heatmap infer (i) the SALI value is high, and (ii) there is a high difference in the activity values of the compound pairs.

3.3.2　Structure-Activity Relationship Index (SARI)

QSAR predictions are based on the similarity principle, assuming a continuous structure-activity relationship (SAR) as reflected by rolling hills with gentle slopes in the SAR landscape, implying continuous changes in the potency with gradual structural changes or increasing diversity. However, the presence of activity cliffs may lead to the occurrence of rugged SAR landscapes, resulting in discontinuous SAR and errors in QSAR predictions [16]. The SAR landscapes may also be heterogeneous in nature, with the occurrence of both continuous and discontinuous regions.

Peltason and Bajorath [17] developed a structure-activity relationship index (SARI) based on the systematic correlation of 2D structural similarity and compound activity. This index can categorize the structure-activity relationship (SAR) into several types: continuous, discontinuous, and heterogeneous. The global SARI has two distinct components: a continuity score or $score_{cont}$ (potency-weighted structural diversity within a class of active compounds, which is characteristic of continuous SARs) and a discontinuity score or $score_{discont}$ (average potency differences (cutoff 1) for pairs of similar ligands exceeding a predefined similarity (Tanimoto coefficient > 0.65), which is a major characteristic of discontinuous SARs and the presence of activity cliffs). These scores are standardized and then transformed into the range of 0–1 by calculating the cumulative probability distribution for each score under the assumption of a normal distribution.

$$SARI = 1/2\left(score_{cont} + \left(1 - score_{discont}\right)\right)$$

Hence, the SARI values take the range 0–1. A SARI value close to zero indicates a discontinuous SAR, a value close to 1 indicates a continuous SAR, and the intermediate values indicate a heterogeneous SAR.

Bajorath's group [18] also introduced a local version of SARI integrated with network-like similarity graphs (NSGs) to identify key compounds that are significant determinants of SAR characteristics, including activity cliffs. In the local version, the discontinuity score is computed individually for each compound, considering all pairs of the compound with other molecules in the activity class, disregarding the potency difference threshold of 1. NSGs depict the similarity and potency relationships (Tanimoto coefficient >0.65) within an activity class by visualizing the compounds as nodes (size scaled to the local discontinuity score), and the similarity relationships as edges. Compounds with larger nodes can be identified as activity cliffs. The visualization of the SAR landscape with NSGs, where node size is scaled according to local discontinuity scores, has also been used in several recent studies [19–22] for identifying activity cliffs and defining improved applicability domain measures for machine learning models.

3.3.3   Structure-Activity Similarity (SAS) Map

Adding the biological potency information as an additional dimension to the chemical space generates the activity landscape, an analysis of which can help detect the activity cliffs. Maggiora and colleagues [23] presented consensus Structure Activity Similarity (SAS) maps, which were developed using the mean of the similarity values of several representations. The objective was to reduce the dependence of the activity landscape on the specific molecular representation. These approaches can identify the activity cliffs in the landscapes generated from multiple molecular representations. In a typical SAS map, the molecular and potency similarities are plotted along the X- and Y-axes, generating four distinct regions (Fig. 3.2). The first region corresponds to low structural similarity instead of high potency similarity (scaffold or side chain hopping). The second region indicates continuous SARs with high values of structural similarity and potency similarity. The third region corresponds to compounds with limited structural and potency similarities. The fourth region corresponds to activity cliffs showing high structural similarity despite poor potency similarity. The authors also showed the use of structure-property-activity similarity in analyzing SARs.

The group of Jose L. Medina-Franco developed an activity landscape plotter that uses the SALI index for the visualization of ACs [24]. This SAS map [25] illustrates the various zones where ACs, similarity cliffs, and suitable data points for QSAR modeling can be identified. The X-axis represents the structural similarity, while the

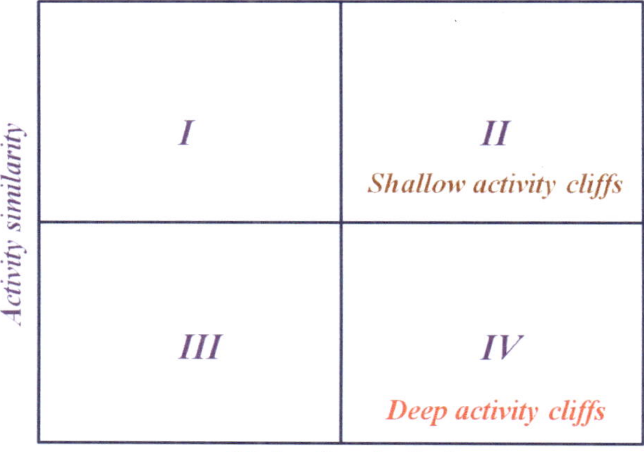

Fig. 3.2 A sample SAS map showing different regions

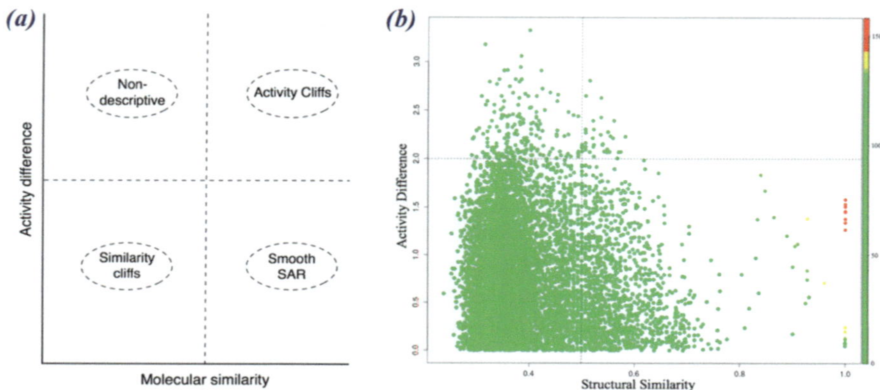

Fig. 3.3 SAS plots (**a**) representing the meaning of different zones, and (**b**) a sample SAS plot generated from the Activity Landscape Plotter. The data points are colored according to SALI values. (Reprinted with permission from the work of Gonzalez-Medina et al. [24])

Y-axis represents the activity difference. The color intensity of the data points is based on the SALI values, the most active compound in the pair, or density. User-defined thresholds are derived for the activity difference and similarity values, and the map is divided into four quadrants. The points lying in the first quadrant (top-right) are typically for ACs, as this characterizes a high activity difference, but a high level of similarity between analog pairs. The points in the fourth quadrant (bottom-right) can be labelled as those that most effectively obey the similarity principle (low activity difference and high similarity between analog pairs), which can be used to develop QSAR models smoothly. The points in the third quadrant (bottom-left) can be labeled as similarity cliffs, as these analog pairs have low similarity, although they exhibit similar activity values (with a low activity difference). Points lying in the second quadrant can be termed non-descriptive. A representation of this concept has been provided in Fig. 3.3a, while Fig. 3.3b represents a sample SAS map generated by the activity landscape plotter.

Various research groups have adopted the SALI index and identified ACs using the SAS plots. Pore and Roy [26] developed an AC plot, where the X-axis represents the Laplacian Kernel similarity, the Y-axis represents the activity difference, and the data points are colored according to SALI values. Pandey and Roy [27] developed a similar AC plot and identified ACs. The plot suggests that the dataset is quite good, with a minimal number of ACs.

3.3.4   The Banerjee-Roy Similarity Coefficients (s_m^1 and s_m^2)

Developed by Banerjee and Roy [28], the coefficients s_m^1 and s_m^2 can efficiently identify ACs. As per the similarity principle, if we analyze the nearest neighbors for a particular query compound, it is expected that (i) the first nearest neighbor should have the same class label as the target compound, and (ii) the average similarity of the nearest neighboring members, belonging to the same class as the target compound, should be high. Adhering to these simple observations, s_m^1 and s_m^2 can be defined as per Eqs. 3.2 and 3.3.

$$s_m^1 = \frac{MaxPos - MaxNeg}{argmax\left(MaxPos, MaxNeg\right)} \tag{3.2}$$

$$s_m^2 = \frac{PosAvgSim - NegAvgSim}{Avg.Sim} \tag{3.3}$$

In Eqs. 3.2 and 3.3, *MaxPos* and *MaxNeg* refer to the maximum similarity value to the positive/active and negative/inactive compounds, respectively, while *PosAvgSim* and *NegAvgSim* refer to the average similarity value of only the positive and the negative nearest neighboring members. *Avg. Sim* is the overall average similarity of the nearest neighboring compounds. For an active compound, it is expected that the *MaxPos* value will be higher than *MaxNeg*, since the nearest neighbor of an active compound should ideally be an active compound, which exhibits a higher level of similarity as compared to the nearest inactive neighbor. Similarly, for an inactive compound, it is expected that the *MaxNeg* value will be higher than *MaxPos*. Therefore, as per Eq. 3.2, the value of s_m will be positive for active compounds and negative for inactive compounds. However, there may be instances where the value of s_m^1 may be negative for an active compound or positive for an inactive compound. This corresponds to the violation of the first assumption, i.e., the first nearest neighbor should ideally be of the same class as the query compound. Similarly, based on consideration of average similarities of the actives and inactives among members of the nearest neighbors, it is expected that the value of *PosAvgSim* will be higher than the value of *NegAvgSim* for an active query compound, and vice versa. Therefore, the value of s_m^2 (Eq. 3.3) should be positive for active compounds and negative for inactive compounds. Again, some compounds may violate the second assumption, which expects an active query compound to have a higher similarity to active/positive close congeners, and vice versa. This results in the generation of positive s_m^1 and s_m^2 values for negative compounds or negative s_m^1 and s_m^2 values for positive compounds. A query compound, which violates both the above-mentioned assumptions, can be termed an activity cliff. On viewing the nearest neighboring compounds for such ACs, it can be observed that the dominance of the opposite class data is prominent, which infers that the structure of the query compound is more similar to the members of the opposite class.

3.3.5   The ARKA Framework

The ARKA framework is a recently reported supervised dimensionality reduction framework that can be used to identify ACs [2]. The theory behind this approach is that it identifies both positively and negatively contributing features, encapsulating their information into two distinct ARKA descriptors. While ARKA_1 encodes the information of the positively contributing features, ARKA_2 encodes the information of the negatively contributing features.

The descriptors in the training set are first normalized to a range of 0–1 to bring them under a uniform scale. For each descriptor, the mean normalized values for the active and inactive class data points are computed. If the mean is higher for the active class data points, it infers that the descriptors contribute positively. Similarly, if the mean is higher for the inactive class data points, it infers that the descriptor contributes negatively. From this data, the mean differences for each descriptor are computed, and subsequently, the different descriptors are grouped according to their contributions to ARKA_1 and ARKA_2. However, for each group, different descriptors have varying magnitudes of mean difference values, which is why it is essential to assign weightage to each descriptor in a group. A descriptor having a higher magnitude of mean difference gets a higher weightage than a descriptor having a comparatively lower magnitude of mean difference in the same class. Thereafter, the training set descriptor matrix is standardized [29], and the ARKA descriptors are computed employing a weighted summation approach using the standardized descriptor values. Therefore, ARKA_1 and ARKA_2 are computed as per Eqs. 3.4 and 3.5.

$$ARKA\_1 = w_1 \times x_1 + w_2 \times x_2 + w_3 \times x_3 \tag{3.4}$$

$$ARKA\_2 = w_4 \times x_4 + w_5 \times x_5 \tag{3.5}$$

Let x_1, x_2, and x_3 be the descriptors that have a positive mean difference value (higher mean in the active class data), and x_4 and x_5 be the descriptors that have a negative mean difference value (higher mean in the negative class data). $w_1 - w_5$ are the corresponding weights.

Therefore, for an active compound, it is expected that it will have a positive ARKA_1 and a negative ARKA_2 value, and an inactive compound should have a negative ARKA_1 and a positive ARKA_2 value. However, if an active compound has a negative ARKA_1 and positive ARKA_2 value or vice versa, then such compounds can be termed ACs. A plot of ARKA_2 vs. ARKA_1 shows the presence of ACs. According to this plot, the positive data points should lie in the fourth quadrant (bottom right) and the negative data points should lie in the second quadrant (top left). However, if negative data points lie in the fourth quadrant or positive data points lie in the second quadrant, such data points can be classified as ACs. The next chapter (Chap. 4) is dedicated to a detailed analysis of the ARKA framework, and therefore, we limit the explanations here.

3.3.6 Machine Learning (ML)-Assisted Prediction of Cliffs

Thus far, we have discussed various cheminformatic indices/descriptors that aid in identifying ACs based on the similarity principle. These indices are independent of the ML modeling algorithm employed, and are solely based on the input feature space and the similarity measure. Modelers can use them to identify ACs before developing models. However, this section presents an approach that utilizes ML models to predict ACs. This is also an important aspect, as ACs are known to misdirect QSAR model development, resulting in higher prediction errors or lower prediction accuracy.

Dablander et al. [30] explored the AC prediction power of various ML models. The authors developed nine QSAR models, employing some of the commonly used machine learning techniques, such as Random Forests, k-Nearest Neighbors, and Multilayer Perceptrons, using three different combinations of feature spaces (i.e., Extended-Connectivity Fingerprints or ECFPs, physicochemical descriptor vectors, and graph isomorphism networks). Using each of the developed models, the authors predicted pairs of similar compounds as ACs or non-ACs. The MMPs were detected using the algorithm presented by Dalke et al. [31], and they were classified into three categories: ACs, non-ACs, and half-ACs. This classification was based on the difference in activity values. If this difference was at least by a factor of 100, then the MMP was labelled as an AC. However, if the difference is below 10, then the MMP was tagged under the non-AC class. The remaining MMPs were labelled half-ACs. In the study, the authors only considered ACs and non-ACs, eliminating the half-ACs, aiming to arrive at a confident binary classification. The authors inferred that the graph isomorphism features were found to be superior to the classical molecular representations. The authors also reported a linear relationship between the QSAR-MAE (Mean Absolute Error) values and the AC-MCC (Matthews Correlation Coefficient). A highly negative Pearson's correlation coefficient (r) was observed between the QSAR-MAE and AC-MCC. With this observation, the authors inferred that as the predictive capability of the model improves, in terms of decreased MAE values, the accuracy in predicting the activity differences among MMPs also improves, leading to improved predictions for ACs.

Tamura et al. [32] used ML and deep learning (DL) models for the large-scale prediction of ACs across 100 activity classes. The authors showed that the prediction accuracy did not scale with the complexity of the methods. For example, deep learning methods showed inferior performance compared to support vector machines, which were also better than the tree-based methods. Surprisingly, the best predictions came from training based on relatively small training sets, considering the specific activity classes. In contrast to conventional ML model development, imbalanced training sets seem more logical for model development, as the number of ACs is generally limited compared to non-ACs. The overall prediction accuracy for the non-AC class may be overestimated for imbalanced sets; therefore, the Matthews Correlation Coefficient, which assigns equal weight to true positives, false positives, true negatives, and false negatives, appears more relevant for evaluating the model's performance. Furthermore, the data leakage issue is particularly important for predictions of ACs: overlapping of compounds between different ACs in training and test sets contributes to the accurate predictions. There is an intricate relationship between data balancing and data leakage for the prediction performance of ACs.

3.4 The Influence of Activity Cliffs on Dataset Modelability and Predictions

Activity cliffs are extreme examples of SAR discontinuity, as minor chemical modifications can lead to significant changes in specific biological activities, resulting in a rugged or canyon-like topology in the activity landscapes. As such molecules do not obey the similarity principle, the QSAR predictions for such molecules may be disappointing [33]. The occurrence of ACs in a dataset may significantly impact its modelability. Good quality data present near the cliff regions may appear to be prediction outliers [33]. Therefore, it may be challenging to develop predictive models from a data set with many activity cliffs. MMPs can avoid several ambiguities associated with continuous similarity measures and thus be used for modeling activity cliffs. Weighted similarity density between a query compound and training compounds, along with inconsistencies in biological activities (discontinuity), has been proposed to define the applicability domain of machine learning models [22], as these metrics have been shown to influence the modelability of the datasets [21]. Prediction outliers often arise due to activity cliff behaviour, which purely linear models cannot address. Upon detecting activity cliffs, additional compounds near the cliffs should be evaluated to ensure that the activity landscapes have been adequately represented [33]. In summary, activity cliffs are highly important in medicinal chemistry, as a proper understanding of the discontinuity may give a significant mechanistic insight, resulting in the synthesis of a new series of potent compounds [34]. By properly analyzing these cliffs, medicinal chemists can understand the complex drug-target interactions more efficiently, leading to improved lead optimization.

References

1. Stumpfe D, Hu H, Bajorath J (2019) Evolving concept of activity cliffs. ACS Omega 4:14360–14368. https://doi.org/10.1021/acsomega.9b02221
2. Banerjee A, Roy K (2024) ARKA: a framework of dimensionality reduction for machine-learning classification modeling, risk assessment, and data gap-filling of sparse environmental toxicity data. Environ Sci Processes Impacts 26:991–1007. https://doi.org/10.1039/D4EM00173G
3. Maggiora G, Vogt M, Stumpfe D, Bajorath J (2014) Molecular similarity in medicinal chemistry. J Med Chem 57:3186–3204. https://doi.org/10.1021/jm401411z
4. Hu Y, Bajorath J (2012) Extending the activity cliff concept: structural categorization of activity cliffs and systematic identification of different types of cliffs in the ChEMBL database. J Chem Inf Model 52:1806–1811. https://doi.org/10.1021/ci300274c
5. Hu X, Hu Y, Vogt M, Stumpfe D, Bajorath J (2012) MMP-cliffs: systematic identification of activity cliffs on the basis of matched molecular pairs. J Chem Inf Model 52:1138–1145. https://doi.org/10.1021/ci3001138
6. Heikamp K, Hu X, Yan A, Bajorath J (2012) Prediction of activity cliffs using support vector machines. J Chem Inf Model 52:2354–2365. https://doi.org/10.1021/ci300306a

7. Stumpfe D, Hu Y, Dimova D, Bajorath J (2014) Recent progress in understanding activity cliffs and their utility in medicinal chemistry. J Med Chem 57:18–28. https://doi.org/10.1021/jm401120g

8. Stumpfe D, Dimova D, Bajorath J (2014) Composition and topology of activity cliff clusters formed by bioactive compounds. J Chem Inf Model 54:451–461. https://doi.org/10.1021/ci400728r

9. Banerjee A, Roy K (2024) The application of chemical similarity measures in an unconventional modeling framework c-RASAR along with dimensionality reduction techniques to a representative hepatotoxicity dataset. Sci Rep 14:20812. https://doi.org/10.1038/s41598-024-71892-4

10. Banerjee A, Roy K (2025) Machine learning assisted classification RASAR modeling for the nephrotoxicity potential of a curated set of orally active drugs. Sci Rep 15:808. https://doi.org/10.1038/s41598-024-85063-y

11. Hu H, Stumpfe D, Bajorath J (2018) Rationalizing the formation of activity cliffs in different compound data sets. ACS Omega 3:7736–7744. https://doi.org/10.1021/acsomega.8b01188

12. Hu H, Stumpfe D, Bajorath J (2019) Second-generation activity cliffs identified on the basis of target set-dependent potency difference criteria. Future Med Chem 11:379–394. https://doi.org/10.4155/fmc-2018-0299

13. Stumpfe D, Bajorath J (2012) Exploring activity cliffs in medicinal chemistry. J Med Chem 55:2932–2942. https://doi.org/10.1021/jm201706b

14. Stumpfe D, Hu H, Bajorath J (2019) Introducing a new category of activity cliffs with chemical modifications at multiple sites and rationalizing contributions of individual substitutions. Bioorg Med Chem 27:3605–3612. https://doi.org/10.1016/j.bmc.2019.06.045

15. Guha R, Van Drie JH (2008) Structure–activity landscape index: identifying and quantifying activity cliffs. J Chem Inf Model 48:646–658. https://doi.org/10.1021/ci7004093

16. Eckert H, Bajorath J (2007) Molecular similarity analysis in virtual screening: foundations, limitations and novel approaches. Drug Discov Today 12:225–233. https://doi.org/10.1016/j.drudis.2007.01.011

17. Peltason L, Bajorath J (2007) SAR index: quantifying the nature of structure-activity relationships. J Med Chem 50:5571–5578. https://doi.org/10.1021/jm0705713

18. Wawer M, Peltason L, Weskamp N, Teckentrup A, Bajorath J (2008) Structure-activity relationship anatomy by network-like similarity graphs and local structure-activity relationship indices. J Med Chem 51:6075–6084. https://doi.org/10.1021/jm800867g

19. Wang Z, Chen J, Hong H (2021) Developing QSAR models with defined applicability domains on PPARγ binding affinity using large data sets and machine learning algorithms. Environ Sci Technol 55:6857–6866. https://doi.org/10.1021/acs.est.0c07040

20. Wang H, Wang Z, Chen J, Liu W (2022) Graph attention network model with defined applicability domains for screening PBT chemicals. Environ Sci Technol 56:6774–6785. https://doi.org/10.1021/acs.est.2c00765

21. Liu W, Wang Z, Chen J, Tang W, Wang H (2023) Machine learning model for screening thyroid stimulating hormone receptor agonists based on updated datasets and improved applicability domain metrics. Chem Res Toxicol 36:947–958. https://doi.org/10.1021/acs.chemrestox.3c00074

22. Zhang Y, Liu Y, Liu W, Chen J (2025) Enlarged data sets and innovative applicability domain characterization empower ml models to reliably bridge hERG binding data gaps in diverse chemicals. Chem Res Toxicol 38(9):1460–1471. https://doi.org/10.1021/acs.chemrestox.5c00065

23. Yongye AB, Byler K, Santos R, Martínez-Mayorga K, Maggiora GM, Medina-Franco JL (2011) Consensus models of activity landscapes with multiple chemical, conformer, and property representations. J Chem Inf Model 51:1259–1270. https://doi.org/10.1021/ci200081k

24. Gonzalez-Medina M, Mendez-Lucio O, Medina-Franco JL (2017) Activity landscape plotter: a web-based application for the analysis of structure–activity relationships. J Chem Inf Model 57:397–402. https://doi.org/10.1021/acs.jcim.6b00776

25. Maggiora GM, Shanmugasundaram V (2011) Molecular similarity measures. In: Bajorath J (ed) Chemoinformatics and computational chemical biology. Methods in molecular biology, vol 672. Humana Press, Totowa, NJ. https://doi.org/10.1007/978-1-60761-839-3_2

26. Pore S, Roy K (2024) Insights into pharmacokinetic properties for exposure chemicals: predictive modelling of human plasma fraction unbound (fu) and hepatocyte intrinsic clearance (Clint) data using machine learning. Digit Discov 3:1852–1877. https://doi.org/10.1039/D4DD00082J

27. Pandey SK, Roy K (2025) Hybrid model development through the integration of quantitative read-across (qRA) hypothesis with the QSAR framework: an alternative risk assessment of acute inhalation toxicity testing in rats. Chemosphere 370:143931. https://doi.org/10.1016/j.chemosphere.2024.143931

28. Banerjee A, Roy K (2023) Prediction-inspired intelligent training for the development of classification read-across structure–activity relationship (c-RASAR) models for organic skin sensitizers: assessment of classification error rate from novel similarity coefficients. Chem Res Toxicol 36:1518–1531. https://doi.org/10.1021/acs.chemrestox.3c00155

29. Snedecor GW, Cochran WG (1991) Statistical methods.8th edn. Wiley-Blackwell, Hoboken

30. Dablander M, Hanser T, Lambiotte R, Morris GM (2023) Exploring QSAR models for activity-cliff prediction. J Cheminform 15:47. https://doi.org/10.1186/s13321-023-00708-w

31. Dalke A, Hert J, Kramer C (2018) mmpdb: an open-source matched molecular pair platform for large multiproperty data sets. J Chem Inf Model 58:902–910. https://doi.org/10.1021/acs.jcim.8b00173

32. Tamura S, Miyao T, Bajorath J (2023) Large-scale prediction of activity cliffs using machine and deep learning methods of increasing complexity. J Cheminform 15:4. https://doi.org/10.1186/s13321-022-00676-7

33. Maggiora GM (2006) On outliers and activity cliffs—why QSAR often disappoints. J Chem Inf Model 46:1535. https://doi.org/10.1021/ci060117s

34. Hu X, Liu G, Zhao Y, Zhang H (2025) Activity cliff-aware reinforcement learning for *de novo* drug design. J Cheminform 17:54. https://doi.org/10.1186/s13321-025-01006-3

Chapter 4
The Arithmetic Residuals in K-Groups Analysis (ARKA) for the Detection of Activity Cliffs

Abstract The Arithmetic Residuals in K-groups Analysis (ARKA) framework has recently been proposed as a supervised dimensionality reduction technique, which has been shown to identify the activity cliffs. In this approach, the training data points are partitioned into active and inactive classes; the descriptors contributing more to each class are identified using the most discriminating features approach. Finally, the weightage of each descriptor for the respective classes is computed, and the entire descriptor matrix is reduced to two dimensions: ARKA_1 and ARKA_2. The scatter plots of the data points in the reduced dimension space can identify the activity cliffs. The descriptors ARKA_1 and ARKA_2 have also been shown to model binary classification of small data sets. More recently, a multi-class ARKA framework has been proposed, considering the contribution of different QSAR descriptors to different activity ranges. This approach has been useful in regression-based model development, particularly when integrated with the similarity-based quantitative read-across structure-activity relationship (q-RASAR) modeling.

Keywords Activity cliffs · ARKA · Dimensionality reduction · Predictions · QSAR · Modelability

4.1 Introduction

Activity cliffs (ACs) are pairs of compounds that show a large difference in their potency while maintaining a high level of chemical similarity [1]. Such compounds, when present in a dataset, misdirect the modeling algorithm and hinder the proper extraction of the relationship between the chemical features and the target endpoint. Therefore, the efficient identification of activity cliffs in a dataset presents a significant challenge to QSAR modelers. Although there have been sporadic attempts to identify ACs based on cheminformatic concepts [2–4], the recently developed supervised dimensionality reduction approach—the Arithmetic Residuals in K-groups Analysis (ARKA)—shows much promise in not only identifying ACs, but

K. Roy, A. Banerjee, *Activity Cliffs*, SpringerBriefs in Molecular Science, https://doi.org/10.1007/978-3-032-10081-8_4

also capturing the response range-specific information of the contributing features [5].

Dimensionality reduction is an integral aspect in cheminformatics that aims to reduce the number of modeling features without losing a significant amount of chemical information. Typically, Principal Component Analysis (PCA) [6] and t-distributed Stochastic Neighbor Embedding (t-SNE) [7] are among the most widely used dimensionality reduction techniques that help to analyze dataset diversity and data visualization. However, due to their "unsupervised" nature, they are not associated with the detection and identification of ACs. On the other hand, ARKA is a supervised dimensionality reduction technique that effectively reduces feature dimensions by utilizing information from the training set's experimental response values. Although the algorithm adopted in this approach is aimed towards dimensionality reduction, it has been observed that the lower-dimensional feature space thus generated can help in the identification of ACs. The detailed theory and the methodological aspects will be explained later.

4.2 The "Curse of Dimensionality" in Small Dataset Modeling

The Curse of Dimensionality is a special problem associated with cheminformatics where the consideration of a large number of features hinders the robustness and generalizability of the model [8]. From the viewpoint of a classical QSAR modeler, the ratio of the number of descriptors to the number of training data points should not exceed 1:5 [9]. However, modern QSAR research based on Machine Learning (ML) often does not consider this ratio. Considering a large feature pool to train a limited number of compounds may mislead the modeling algorithm into identifying meaningful patterns in the data, as the model tries to forcefully identify feature contributions, which apparently do not relate to the target endpoint. In such cases, the model learns the "noise" associated with the redundant/insignificantly contributing features, and therefore fails to generalize properly with unseen data. Such models are deemed overfitted and are typically characterized by a sharp decrease in the cross-validation statistics compared to the over-optimistic fitting statistics. Figure 4.1 demonstrates the effect of considering unnecessary data dimensionality.

If we adhere to the Topliss and Costello's 5:1 ratio [9], this phenomenon may happen in mainly two different cases. Firstly, when we consider a large number of modeling features, and secondly, when we consider "too few" training data points. The second case is mostly associated with small dataset modeling, where there is a scarcity of experimentally determined endpoint values of chemicals. With the increasing need for data-gap filling, small dataset modeling is adopted, especially in fields like nanotoxicology and nano-property predictions [10]. While small dataset modeling requires a perfect balance between the training data points and the just-right amount of modeling descriptors, this can be bypassed by adopting

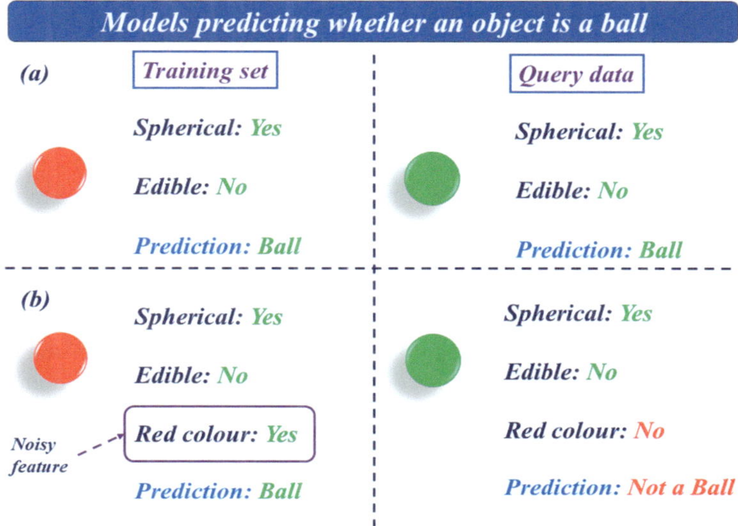

Fig. 4.1 Role of noisy features towards misclassification. (**a**) A model trained with just the right amount of feature space, and correctly classifies the query data. (**b**) A model trained with a noisy feature generates incorrect predictions for the query data

Read-Across approaches [11] that are non-statistical and do not require fulfilling a sufficient degree of freedom of QSAR models. However, the quantitative interpretability of most Read-Across approaches suffers as the relative contributions of descriptors are not expressed [12]. Therefore, an alternative approach can be to adopt dimensionality reduction techniques for model development, which (i) preserves interpretability and (ii) improves the statistical degree of freedom by lowering the number of modeling descriptors.

4.3   The ARKA Algorithm for the Supervised Dimensionality Reduction

As discussed above, small dataset modeling can be very challenging as it requires capturing essential information, but the limited number of training compounds does not allow for considering a greater number of modeling descriptors. To avoid statistical complexity and to retain the required feature space, dimensionality reduction approaches can be used. The ARKA framework is a supervised dimensionality reduction technique that is fundamentally based on the following theories: firstly, to identify the relative contribution of descriptors towards data points belonging to different class labels, and secondly, to identify their degree of contributions among the same classes of data points, considering the effects of various other descriptors.

The information of the contributing descriptors is stored in two different ARKA descriptors (ARKA_1 and ARKA_2) depending on their contributions. Information for the positively contributing descriptors is stored in ARKA_1, and the information for the negatively contributing descriptors is stored in ARKA_2. The identification of the positive and negative contributions of descriptors has been detailed in the following paragraph, where we discuss the algorithmic details.

Unlike other dimensionality reduction techniques, such as PCA, the algorithm of the ARKA framework involves simple mathematical concepts that are computationally inexpensive. Inspired by the pictorial representation of neural networks presented in Roy et al. [13], the ARKA framework follows a similar pathway, consisting of an input set of descriptors, a hidden layer where calculations related to dimensionality reduction occur, and a final output layer that presents the lower-dimensional feature matrix. At first, the training set descriptor matrix is used for the computation of descriptor contributions and weightages. Considering a training set with binary classification labels (active/inactive or toxic/non-toxic), each descriptor is normalized from 0 to 1 using Eq. 4.1 [14].

$$n_{ij} = \frac{Value_{ij} - Min_i}{Max_i - Min_i} \tag{4.1}$$

In Eq. 4.1, n_{ij} represents the normalized value of the descriptor i for the jth data point. Min_i represents the minimum value of the ith descriptor in the training set, while Max_i represents the maximum value of the ith descriptor in the training set. This is an important pre-processing step that aims to unify the range of different descriptors for their proper mathematical analysis.

After normalization of the training set feature matrix, the mean values of each descriptor are calculated for the active (positive class mean) and inactive (negative class mean) classes of data. In a general sense, if the positive class mean value is greater than the negative class mean value, it can be inferred that the particular descriptor is positively contributing, since it has higher values in the active class data. An opposite observation is also possible, where the negative class mean is higher than the positive class mean, which infers that the descriptor is negatively contributing. The mean difference values are then calculated for each descriptor, where the magnitude of the mean difference governs the level of weightage a particular descriptor receives. Although "K" in ARKA stands for K-groups, the purpose of activity cliff analysis and classification modeling restricts the value of K to 2, since this depicts that the data can only be split into two different classes for binary classification problems. However, for regression problems, the value of K can be increased, leading to the definition of "multiclass" ARKA, which will be discussed in detail in dedicated sections.

After the computation of the mean difference values (positive class mean − negative class mean) for each descriptor, they are then segregated into two distinct classes/groups. The first group represents the descriptors having a positive mean difference value, while the second group represents the descriptors having a negative mean difference value. Among the descriptors constituting a particular group, higher weightages are given to descriptors that possess a higher mean difference value, and vice versa. The weightages are defined as in Eq. 4.2.

$$Weightage\ of\ a\ descriptor\ in\ a\ group = \frac{Mean\ difference\ value\ of\ a\ descriptor}{\sum Mean\ difference\ values\ of\ all\ the\ descriptors\ in\ the\ particular\ group}$$

(4.2)

Using Eq. 4.2, weightages are given to all the descriptors constituting a particular group. This is performed twice for the two different groups. It should be noted that while computing the weightage for a particular descriptor in a group, the mean difference values of descriptors to other groups are never considered.

Once the weightages are assigned to the descriptors, the training and test set descriptor matrices are standardized using the formula in Eq. 4.3 [14]. Standardization is a data transformation technique where the descriptor columns have a zero mean and a unit standard deviation.

$$n_{ij} = \frac{x_{ij} - \mu_i}{\sigma_i}$$

(4.3)

In Eq. 4.3, n_{ij} is the standardized value of the descriptor i for the data point j, x_{ij} is the actual value of the descriptor i for the data point j, μ_i is the mean value of the descriptor i in the training set, and σ_i is the standard deviation of the descriptor i in the training set. After the standardization of the training and test set descriptor matrices, ARKA_1 and ARKA_2 are computed based on a weighted summation strategy. Considering that there are five different descriptors x_1 through x_5 in the input feature matrix, and it has been observed that descriptors x_1, x_2, and x_3 belong to group 1, while x_4 and x_5 belong to group 2, the computation of ARKA_1 and ARKA_2 has been shown in Eqs. 4.4 and 4.5.

$$ARKA\_1 = w_1 \times x_1 + w_2 \times x_2 + w_3 \times x_3$$

(4.4)

$$ARKA\_2 = w_4 \times x_4 + w_5 \times x_5$$

(4.5)

In Eqs. 4.4 and 4.5, x_1, ... , x_5 are the five different descriptors, and w_1, ... , w_5 are the corresponding weightages. ARKA_1 and ARKA_2 are the reduced feature dimensions that contain information from all five different descriptors based on their level of contributions towards a particular class of data points.

For the ease of users and to avoid potential computational errors, a simple Java-based tool is freely available from https://sites.google.com/jadavpuruniversity.in/dtc-lab-software/arithmetic-residuals-in-k-groups-analysis-arka, which quickly computes ARKA descriptors for both the training and test sets.

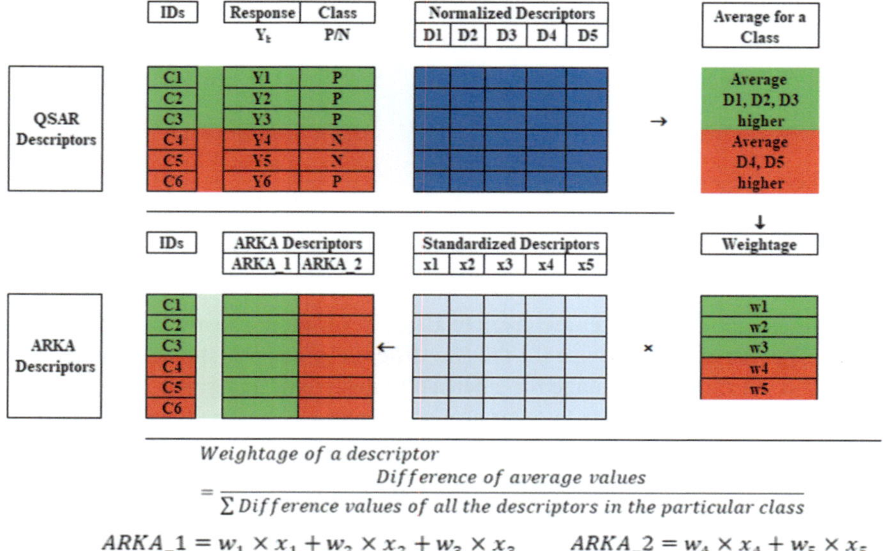

$$\text{Weightage of a descriptor} = \frac{\text{Difference of average values}}{\sum \text{Difference values of all the descriptors in the particular class}}$$

$$ARKA\_1 = w_1 \times x_1 + w_2 \times x_2 + w_3 \times x_3 \qquad ARKA\_2 = w_4 \times x_4 + w_5 \times x_5$$

Fig. 4.2 The algorithm for the computation of ARKA descriptors. (Reprinted with permission from Banerjee A, Roy K (2024) Environ Sci Processes Impacts 26: 991–1007)

The complete algorithm for the computation of the ARKA descriptors has been shown in Fig. 4.2.

4.4   ARKA Descriptors as Identifiers of Activity Cliffs

Activity cliff identification and analysis have been a long-standing challenge for QSAR modelers. This is attributed to the fact that ACs do not obey the similarity principle that states similar compounds should show similar activities. Such data points mislead the modeling algorithm in the identification of meaningful patterns in the training data. Therefore, it becomes imperative to assess the modelability of the dataset and to identify potential ACs before QSAR model development. ARKA descriptors, due to their supervised nature, can help in the identification of ACs.

As per the algorithm for the computation of ARKA descriptors, where the positively contributing descriptors are grouped into ARKA_1 and the negatively contributing descriptors are grouped into ARKA_2, it is expected that for active compounds, ARKA_1 should have a positive value and ARKA_2 should have a negative value, and vice versa for inactive compounds. However, this is not always the case. Figure 4.3 presents the theoretical distribution of different data points in the ARKA descriptor space, and the corresponding interpretation for data points lying in each quadrant.

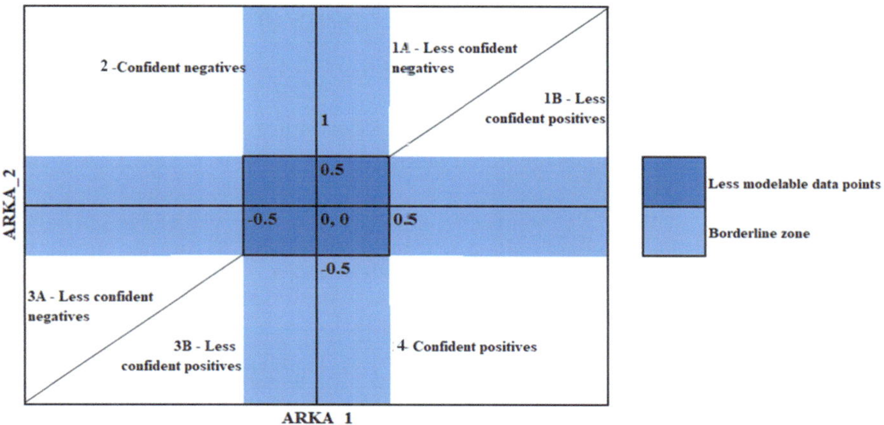

Fig. 4.3 ARKA_2 vs. ARKA_1 plot showing the regions of confident and less-confident predictions. (Reprinted with permission from Banerjee A, Roy K (2024) Environ Sci Processes Impacts 26: 991–1007)

The ARKA_2 vs. ARKA_1 plot shows confident and less confident data points, and is an indicator of the modelability of a dataset. This plot is divided into four quadrants, each possessing a defined characteristic.

Quadrants 2 and 4 (top left and bottom right, respectively) are the quadrants where we can expect confident data points. As stated previously, for positive/active data points, it is expected that they should ideally have a positive ARKA_1 and a negative ARKA_2 value, which positions them in the fourth quadrant (bottom right) of the plot. Similarly, for negative/inactive data points, ARKA_1 should ideally have a negative value, and ARKA_2 should ideally have a positive value. Such compounds position themselves at the second quadrant (top left) of the plot. Data points ideally following this criterion can be termed "confident". However, this ideal situation is not universal. There may be instances where positive compounds have a negative ARKA_1 and a positive ARKA_2, positioning them in the second quadrant, or negative compounds having a positive ARKA_1 and a negative ARKA_2, making them members of the fourth quadrant. Such compounds are "confident" and appear to be located in the opposite region. These compounds, i.e., negative compounds in the confident positive region (fourth quadrant) and positive compounds in the confident negative region (second quadrant), can be termed activity cliffs. The other data points lying in the first and third quadrants are less confident. It should be noted that a buffer zone of ±0.5 from the axes should be used to restrict the stringency of AC identification. Theoretically, data points lying close to the axes, or more specifically, near the origin, have reduced confidence. This confidence increases as the absolute values of ARKA_1 and ARKA_2 increase. Banerjee and Roy extended this by computing the Euclidean Distance of the data points from the origin, which serves as a measure of confidence in the identification of ACs [15].

4.5   The Application of ARKA Descriptors in Identifying Activity Cliffs

Different research groups have employed ARKA descriptors for identifying activity cliffs. Some relevant examples are noted below.

4.5.1   The Application of Similarity Measures to a Hepatotoxicity Data Set

A binary hepatotoxicity data set was modeled [16] using a classification Read-Across Structure-Activity Relationship (c-RASAR) modeling approach that uses read-across-derived similarity and error-based descriptors into a statistical modeling framework. While the external predictivity of the final linear discriminant analysis (LDA)-based c-RASAR model was shown to be much better than that of the corresponding QSAR model, the identification of activity cliffs using the ARKA approach demonstrated that the use of similarity descriptors led to enhanced modelability of the data set due to the presence of a lower number of data points in the borderline and less modelable zones. Five compounds in the training set and four compounds from the test set were identified as activity cliffs. Among these, bupivacaine is a hepatotoxic compound that exhibits high structural similarity to mepivacaine, a non-hepatotoxic compound. Again, dienestrol is a non-hepatotoxic compound showing a high structural similarity to hepatotoxic phenol. The identification of activity cliffs may also help explain wrong instances of predictions from otherwise valid predictive models.

4.5.2   Nephrotoxicity Potential of Orally Active Drugs

The nephrotoxicity potential of a curated list of orally active drugs was modeled [15] using c-RASAR, demonstrating the superior performance of an LDA-based c-RASAR model compared to the corresponding QSAR model. The application of the ARKA approach identified activity cliffs, such as glipizide (non-nephrotoxic) and darunavir (non-nephrotoxic). Eight out of ten closest congeners of glipizide belong to the nephrotoxic class, while all of the ten close source congeners of darunavir are nephrotoxic, explaining their activity cliff behavior. The confidence in the activity cliff behavior may also be judged based on the distance of the query compound from the origin, provided that positive compounds are present in the second quadrant and negative compounds are present in the fourth quadrant. The combination of the ARKA approach with the similarity-based RASAR approach may easily explain several model-derived wrong prediction outcomes.

4.5.3   Combined Toxicity and Interactions of Antibiotic and Fungicide Mixtures

Qin et al. [17] developed machine learning-based classification and regression models for the toxicities of 75 binary mixtures of antibiotics and fungicides against *Auxenochlorella pyrenoidosa*, a species of green microalgae, as an indicator of aquatic toxicity. The authors demonstrated that a random forest (RF) classifier exhibited high accuracy in distinguishing the synergistic and antagonistic mixtures. The authors used the five variables appearing in the RF classification model to calculate ARKA descriptors for both the training and test sets. From the scatter plots of ARKA_2 vs. ARKA_1 separately for the training and test sets, the authors identified one mixture in the training set and three mixtures in the test set as the activity cliffs.

4.5.4   Toxicity of Ionic Liquids on Leukemia Rat Cell Line IPC-81, Escherichia coli (E. coli), and Acetylcholinesterase (AChE)

Shan et al. [18] developed random forest and gradient boosted decision tree models using three different molecular representations, including molecular descriptors, molecular fingerprints, and molecular identifiers, to assess the toxicity of ionic liquids on the leukemia rat cell line IPC-81, *Escherichia coli* (*E. coli*), and acetylcholinesterase (AChE). The authors demonstrated that models derived using molecular descriptors performed more effectively in modeling the three endpoints. The authors also attempted to identify activity cliffs using the ARKA_2 vs. ARKA_1 plots for the training and test sets in the case of the leukemia rat cell line IPC-81 data set; however, they did not properly indicate the cliffs in the ARKA plot.

4.5.5   Ames Mutagenicity Predictions Based on 2D Descriptors

Borah and Nagamani [19] developed machine learning models (random forest, XGBoost, and CatBoost) using 2D descriptors (RDKit, Mordred, and CDK) for Ames mutagenicity predictions. The XGBoost model was shown to perform the best among the reported models. The authors could identify several activity cliffs using the ARKA approach, starting from the initially selected 23 descriptors. Among the identified cliffs, the authors explained the behaviour of two cliffs, one from the training set and another from the test set, considering their respective close congeners.

4.5.6   Toxicity of Organic Chemicals to Three Trout Species

Kar and Gallagher [20] developed QSAR and quantitative read-across structure-activity relationship (q-RASAR) models for the acute toxicity of chemicals against three North American trout species: *Oncorhynchus clarkii, Salvelinus fontinalis, and Salvelinus namaycush.* The authors demonstrated that q-RASAR models outperformed QSAR models, yielding better internal and external model statistics. The authors used the ARKA approach to identify the activity cliffs; however, none of the data points emerged as activity cliffs from the analysis.

4.5.7   Fungicidal Activity of Binary Mixtures on Macrophomina phaseolina

Rahimi-Soujeh et al. [21] modeled the fungicidal activity of binary mixtures on *Macrophomina phaseolina* using QSAR and read-across approaches. They demonstrated that Support Vector Regression (SVR) and Gaussian Process Regression (GPR) models outperformed other models in terms of predictive capacity. The authors applied the ARKA_2 vs. ARKA-1 plot to identify activity cliffs. However, the authors opined that none of the mixtures exhibited activity cliffs, except for one mixture, *Teb-Pro R3*, which appeared as a marginal activity cliff.

4.5.8   Toxicity of Ionic Liquids Against the Leukemia Rat Cell Line (IPC-81)

Abdellatif et al. [22] modeled the toxicity of ionic liquids against the leukemia rat cell line (IPC-81) using the dragonfly algorithm (DA)-optimized support vector machine (SVM) model. The authors identified activity cliffs in the training and test sets using the ARKA approach. The authors comment that the cliffs represent areas where the model may struggle to classify compounds correctly.

4.5.9   Acute Toxicity of Pesticides to the Sheepshead Minnow

Sun et al. [23] developed a random forest model to predict the acute toxicity of pesticides to the Sheepshead minnow. While SHAP analysis explored the relative importance of different features, the analysis of ARKA descriptors indicated the absence of any activity cliffs.

4.5.10   Rat and Mouse Acute Oral Toxicity of Antimicrobial Agents

Banjare et al. [24] investigated the acute toxicity of antimicrobials in rats and mice using the QSAR approach, and also explored interspecies correlation analysis. The authors developed multiple linear regression models with a genetic algorithm as the feature selection technique. They employed the leverage approach to define the applicability domain and the ARKA approach to identify prediction outliers.

4.5.11   Mixture Toxicity of Azole fungicides Toward Algae

Qin et al. [25] applied 12 different machine learning algorithms to model the mixture toxicity of azole fungicides toward algae. The authors showed that a consensus model combining SVM and RF algorithms outperformed the individual models in the quality of predictions. The authors attempted to identify activity cliffs and less confident data points from the ARKA_2 vs. ARKA_1 plots; however, they selected compounds from the incorrect quadrants (first and third quadrants) in their report.

4.5.12   Soil Adsorption Coefficient ($logK_{OC}$) Prediction

Yang et al. [26] developed quantitative structure-property relationship (QSPR) models using a dataset comprising 1477 experimental $logK_{OC}$ values and seven machine learning algorithms. The authors selected and reported random forest, gradient boosting decision tree, and XGBoost models as the best-performing models. For different models, the authors identified activity cliffs according to the ARKA approach.

4.5.13   Respiratory Toxicity of Environmental Pollutants

Li et al. [27] developed a deep learning model for the respiratory toxicity of environmental pollutants and identified the structural alerts associated with the respiratory toxicity. The authors applied the ARKA approach to identify the activity cliffs for various respiratory toxicity endpoints. In case of asthma, pulmonary fibrosis, and respiratory infection, no active cliffs were detected. In contrast, among the remaining endpoints (pneumonia, pulmonary edema, pulmonary embolism, pulmonary arterial hypertension, bronchospasm, bronchitis, and respiratory toxicity), a limited number of active cliffs could be detected.

4.5.14   Placental Barrier-Permeable Contaminants

Gao et al. [28] developed a multifusion deep learning (DL) model incorporating a graph convolutional network for the accurate and rapid screening of chemicals that permeate the placental barrier. The authors found the presence of a limited number of activity cliffs (negative and positive data points distributed in the fourth and second quadrants, respectively), indicating high confidence and modelability of the dataset.

4.5.15   Bioactivity Prediction of Per- and Polyfluoroalkyl Substances (PFAS)

Yin et al. [29] established machine learning QSAR models using quantitative molecular surface analysis (QMSA) of molecular electrostatic potential. The authors identified important descriptors through the SHAP analysis and analyzed the presence of activity cliffs through the ARKA approach.

4.5.16   QSAR of Multigeneration EGFR Inhibitors

Sun et al. [30] developed a multitask deep neural network (MT-DNN) QSAR model for predicting the bioactivities of multigeneration EGFR inhibitors. In this work, the analysis of activity cliffs was done through the computation of ARKA descriptors.

4.5.17   Toxicity of Diverse Industrial Chemicals Towards Salmon Species

Bhattacharyya et al. [31] developed QSAR and q-RASAR models for the toxicity of diverse industrial chemicals against three different salmon species: *Salmo salar,* *Oncorhynchus kisutch,* and *Oncorhynchus tshawytscha.* However, the application of the ARKA approach did not identify any activity cliffs in the data sets.

4.5.18   Algal Toxicity of Chemicals

Yu [32] modeled the toxicity of chemicals against green alga *Raphidocelis subcapitata* using a random forest classifier developed from quantum chemical descriptors including atomic charges, multipole moments, frontier orbital energies, kinetic energies, polarizability, and thermochemistry properties. The author could identify several activity cliffs in the data set using the ARKA approach.

4.5.19   Neurotoxicity of Environmentally Relevant Chemicals

Hao et al. [33] used three molecular representation methods, including molecular fingerprints, molecular descriptors, and molecular graphs. They applied six traditional machine learning algorithms and two deep learning approaches for the prediction of neurotoxicity of environmentally relevant chemicals. These authors used the ARKA approach for the identification of activity cliffs.

4.5.20   Ecological Risk Assessment of Phenylurea Herbicides in Aquatic Environments

Wei et al. [34] developed QSAR models to predict the environmental risk limit (HC5) of phenylurea herbicides. A random forest model was shown to outperform an MLR model. However, no compound was found to be an activity cliff according to the ARKA method.

4.5.21   Activity Coefficient of Aqueous Ionic Liquids

Adhab et al. [35] developed an artificial neural network (ANN)-based quantitative structure-property relationship model using group contributions to estimate the mean ionic activity coefficient of aqueous quaternary ammonium salts. The authors considered critical temperature (Tc), critical volume (Vc), acentric factor (ω), molality, and molecular weight in the input layer. The resulting model showed a robust model performance. The authors identified property cliffs and less confident data points using the ARKA approach.

4.5.22   Inhibition of 5α-Reductase Type 1 by Halogenated Disinfectants

Halogenated disinfectants modulate neurosteroidogenesis through the inhibition of 5α-reductase type 1. Fang et al. [36] explored the structure-activity relationships of these compounds and utilized the ARKA approach for the identification of activity cliffs and cross-species correlation assessment.

4.6 ARKA for Small Dataset Modeling

As stated previously, modeling small datasets can be challenging due to the limited number of training data points. From a cheminformatician's viewpoint, the selected feature space should be sufficiently large to capture the essential information from the limited number of training data points, so that the developed model can generalize well with new data. However, due to the statistical complexity, often there is a trade-off between information loss and limiting the model's degree of freedom. A solution to this problem has been the use of dimensionality reduction techniques, which enable the capture of a greater amount of feature space in a limited number of modeled descriptors. However, apart from removing noise and intercorrelation (if present) among the input features, there is always some loss of information when the features are encapsulated into a reduced number of modeling descriptors. A modeler aims to reduce this information loss, which is why in Principal Components Analysis, we typically consider the optimum number of principal components that encodes 95% of the available information [6]. In the case of computing ARKA descriptors, primarily, the input feature matrix is reduced to just two ARKA descriptors (ARKA_1 and ARKA_2). However, the comprehensive study from Banerjee and Roy [5] showed that despite a drastic reduction in the number of modeling features, and potentially suffering from a significant information loss, the ARKA descriptors generate more robust and predictive models as compared to the corresponding conventional QSAR models. This improved performance has also been observed in the case of Read-Across-based predictions, where the ARKA feature matrix generated better predictions than the conventional descriptor matrix. This is an important observation as both the modeling/prediction strategies used the same level of chemical information, yet generated enhanced performances. This can be explained by the inherent feature of the ARKA descriptors that seem to handle noise among descriptors efficiently. Although in this study the authors employed a Machine Learning-based classification modeling approach, the following case studies highlight that the ARKA descriptors can also perform well in a regression modeling framework.

4.6.1 Risk Assessment of Chemicals Against Salmon Species

Bhattacharya et al. [31] developed global models to predict the toxicity of industrial chemicals against *Salmon* species. They used various modeling strategies that encapsulate Read-Across-derived information and ARKA dimensionality reduction. Initially, they developed the standard QSAR model by computing 0-2D descriptors, followed by data splitting, feature selection, and model development. To improve the predictivity of the model, they have developed q-RASAR models, adhering to the methodology enumerated by Banerjee and Roy [37–39]. Additionally, they have developed ARKA models using the two ARKA descriptors outlined in

[5], which had the highest Q^2_{F1} and Q^2_{F2} values among the individual models. Moreover, they have developed a hybrid model that combines QSAR descriptors, RASAR descriptors, and ARKA descriptors. This hybrid model is a five-descriptor model, where three descriptors are RASAR descriptors and one descriptor each for ARKA and QSAR. Additionally, to further amalgamate the advantages of all the different modeling strategies, the authors developed a Partial Least Squares stacking regression, which showed the best performance in terms of robustness and external predictivity. This global stacking model was further used for true external set prediction.

4.6.2   Prediction of Placental Permeability of Organic Sunscreen Products

Sobanska et al. [40] developed models that predict the placental permeability of drug-like degradation products from organic sunscreens. According to the protocol for conventional QSAR modeling, the authors computed descriptors, performed feature selection, and developed QSAR models. To encode information on the similarity to nearest neighbors, the RASAR descriptors were calculated, and q-RASAR models were developed, adhering to the "best practices" of q-RASAR model development [37–39]. Additionally, the ARKA descriptors were computed, and bivariate models were generated. Among the different QSAR, ARKA, and q-RASAR models developed using two different feature matrices, the bivariate ARKA model stands out in achieving the highest cross-validation (CV) score and F-value, which results after efficient dimensionality reduction. Additionally, the best CV score, coupled with the highest F-value and the lowest number of modeling descriptors employed (only two ARKA descriptors), inferred that the ARKA model is statistically highly stable.

4.7   The Multiclass ARKA Framework for Improved Regression Models

Thus far, the ARKA descriptors have been effective in a binary classification modeling framework that has graded response values. However, for regression modeling, although there have been successful implementations of the ARKA framework in deriving robust and predictive models [31, 40], it has been observed that for datasets having a significantly large range of response values, merely dividing the data points into two different classes, based on training set experimental response mean, generates a significant loss of chemical information. This can be explained in cases where data points having very high (or low) response values get grouped together with data points whose response values are near the training set mean response. Since "K" in ARKA refers to K-groups, such cases necessitate the splitting of the training dataset into multiple groups. The idea behind partitioning the training data into multiple groups is, firstly, to reduce the response range in each group, and secondly, to capture a particular range-specific contribution of the modeling descriptors (discussed later). This aims to reduce noise in the computation of the ARKA descriptors. Initially, the training set response values are normalized from 0 to 1 using Eq. 4.1. The grouping of data points is carried out in such a way that each group has an equal normalized response range. This range can be defined as $1/K$, where K is the number of groups. As an exemplary case, considering $K = 4$, group 1 ranges from 1 to 0.75, group 2 ranges from 0.75 to 0.5, group 3 ranges from 0.5 to 0.25, and group 4 ranges from 0.25 to 0 (Figs. 4.4 and 4.5). It should be noted that the number of data points in each group may vary, as the group is defined based on the normalized response values. When the training set is partitioned into multiple groups, data points from a particular group are considered positive, while all the other data points are considered negative. This is carried out in "K" different iterations, where each group gets to become the positive class (Fig. 4.4). This process can be related to the K-fold CV strategy, where data points from each fold are held out in K different iterations. In each iteration, the ARKA descriptors are computed considering data points from the positive and negative classes, as outlined in the original paper of the ARKA framework [5], and discussed previously in this chapter. Therefore, for "K"-groups, there are "K" different iterations, resulting in "$2K$" number of ARKA descriptors. Again, this concept is an extension of the original ARKA framework and is entirely based on a reproducible mathematical calculation. For the ease of computation of the users, a simple Java-based tool "Multiclass_ARKA-v1.0" has been developed, which is freely available from https://sites.google.com/jadavpuruniversity.in/dtc-lab-software/arithmetic-residuals-in-k-groups-analysis-arka. This tool takes the training and test set files as inputs, and asks the user to provide the number of groups (K) for splitting the training set. Accordingly, the tool calculates $2K$ number of ARKA descriptors for the training and test sets. It should be noted that the tool allows the user to select up to a certain number of groups, which ensures that there are at least two data points in a particular group, thus making the calculations statistically compliant.

The introductory paper of the Multiclass ARKA framework employed this approach on five different toxicity datasets [41]. Each of these datasets has published QSAR and q-RASAR models, which served as standards to evaluate the performance of the multiclass ARKA descriptors. The authors considered K values of 4 and 6 and suggested that these two groups may be used more frequently. This is because the consideration of the optimal number of groups is dependent on the size of the training dataset, and the range in training set response values. If a lower number of groups is considered, there might not be sufficient encoding of the response range-specific information, as the group interval of the response values increases. On the other hand, considering a higher number of groups results in the computation of an increased number of ARKA descriptors, which can encounter multi-collinearity issues when modelled together with the original QSAR descriptors.

Fig. 4.4 Grouping strategy of the Multiclass ARKA framework considering $K = 4$. In each iteration, a particular class/group is considered positive, while the remaining data points are considered negative. X is a descriptor, and Y represents the normalized response values

After computing the multiclass ARKA descriptors, they were then fused with the originally selected QSAR descriptors, and hybrid ARKA models were developed by applying feature selection to the "hybrid" feature matrix. Additionally, using this hybrid feature matrix, similarity between compounds was defined, and RASAR descriptors were computed [39]. This novel framework not only used the structural and physicochemical features of molecules, but also used the response range-specific information of features to define similarity and compute the RASAR

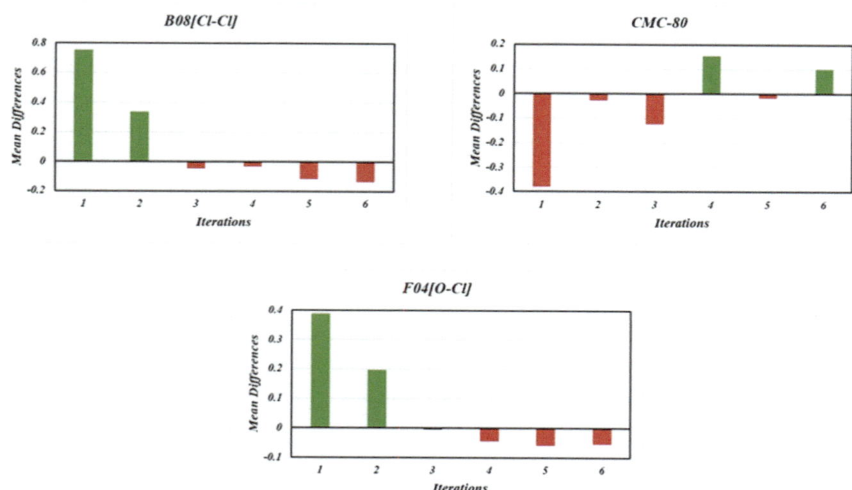

Fig. 4.5 Contributions of different descriptors across different iterations (response ranges). (Reprinted with permission from Banerjee A, Roy K (2025) J Hazard Mater 496: 139302)

descriptors. The computed RASAR descriptors were then fused with the selected hybrid ARKA feature matrix (containing both QSAR and ARKA descriptors), and feature selection algorithms were employed to generate "ARKA-RASAR" models. A multi-criteria decision-making approach was applied to identify the best modelling algorithm (previously reported QSAR, previously reported q-RASAR, Hybrid ARKA, and ARKA-RASAR models), considering various internal and external validation statistics. The results showed that the ARKA-RASAR models (i.e., ARKA integrated into the q-RASAR framework) showed the best performance across the datasets, with improved cross-validation scores and external predictivity. Moreover, this integration seemed to address the concern that conventional q-RASAR models generate slightly inferior internal validation statistics, in some cases, as compared to the corresponding QSAR model. The ARKA-RASAR models not only showed enhancements in internal and CV statistics but also retained the ability to enhance the external predictivity of the conventional q-RASAR models. The Variable Importance Plots (VIPs) of the ARKA-RASAR models for all five different datasets showed that the ARKA descriptors, along with the RASAR descriptors derived from the hybrid feature matrix, showed the highest importance. In contrast, the conventional QSAR descriptors possessed low importance. Additionally, true external validation showed that the ARKA-RASAR models generalize well with unseen data.

4.7.1   Identification of Response Range-Specific Contribution of Descriptors Using Multiclass ARKA: An "Intelligent" Physical Interpretation

Recently, Banerjee and Roy [42] demonstrated how the multiclass ARKA descriptors can identify the relative contributions of different QSAR descriptors across different response ranges. In this paper, the authors modelled the Oral Slope Factor (OSF) and Inhalation Slope Factor (ISF), which are indicators of carcinogenic potential of chemicals, using the Multiclass ARKA framework. Adhering to OECD Principle 5—proposing a mechanistic interpretation if possible—the authors identified the relative contribution of structural and physicochemical features. However, unlike the conventional interpretation techniques that assess the contributions of descriptors across the entire training set chemical space, the multiclass ARKA framework allows for the efficient identification of the varying contributions of a particular descriptor across data points belonging to different response ranges. Figure 4.5 demonstrates the varying contributions of different QSAR descriptors across different iterations/groups of training data. These plots demonstrate the varying mean difference values (positive class mean − negative class mean) for three different descriptors across different iterations. For the descriptor B08[Cl-Cl], it can be observed that the mean difference values are highly positive in the initial iterations, where the data points have the highest observed response values (or highly potent carcinogens). On the other hand, the mean difference values are significantly negative in the later iterations, where the data points have the lowest observed response values (or the least potent carcinogens). In cases where the data points have response values close to the training set mean response, the mean differences are close to zero. This observation suggests that B08[Cl-Cl] exhibits a substantial positive contribution for highly potent carcinogens and a negative contribution for the least potent carcinogens. A similar explanation can also be attributed to the contributions of F04[O-Cl]. However, for the descriptor CMC-80, an opposite trend is observed where the mean difference is highly negative in the initial iterations, and significantly positive in the later iterations. This infers that CMC-80 contributes negatively for highly potent carcinogens, and positively for the least potent carcinogens. This "intelligent" physical interpretation can be made possible by using the multiclass ARKA framework. Additionally, there is a concept of a trade-off between predictivity and interpretability, as complex ML models are black-box in nature. In contrast, interpretable and straightforward models often lack high predictivity towards unseen compounds. The application of the multiclass ARKA framework and the development of ARKA-RASAR models not only enhance robustness and predictivity but also provide an improved version of physical interpretation.

4.7.2   The Applications of the Multiclass ARKA Framework Towards Developing Improved Predictive Models

Since the inception of the multiclass ARKA framework in early 2025 [41], there have been successful applications in the improvement of model qualities. This section discusses these applications and the main findings.

4.7.2.1   Prediction of Carcinogenic Potency of Organic Chemicals

Banerjee and Roy [42] employed a variety of modelling approaches to predict the OSF and ISF, which are indicators of carcinogenic potential, of organic chemicals. At first, conventional QSAR models were developed after following the usual protocol of descriptor computation, data splitting, and feature selection. This was followed by the computation of the RASAR descriptors and the development of conventional q-RASAR models as outlined in [38]. After that, the authors used the selected QSAR descriptor space to compute the multiclass ARKA descriptors. These descriptors were then fused with the selected QSAR descriptors, and hybrid ARKA models were developed after feature selection. This hybrid feature space was used to compute RASAR descriptors, and ARKA-RASAR models were developed as outlined in [41]. Each of the developed models was validated internally and externally. Additionally, various machine learning algorithms were used as stacking regressors, trained on the cross-validated predictions of the QSAR, q-RASAR, hybrid ARKA, and ARKA-RASAR models. A statistical multi-criteria decision-making approach inferred that the ARKA-RASAR and q-RASAR algorithms were the best. This also inferred that the Linear Support Vector Regression and Ridge Regression were the best stacking models for the OSF and ISF data, respectively. A true external set validation inferred that the ARKA-RASAR model generalized well on unseen data.

4.7.2.2   Prediction of Fraction Free Volume (FFV)

Baran and Kloskowski [43] used the multiclass ARKA framework to develop models for FFV. The authors state that the neural network model developed using only the ARKA descriptors showed the best predictivity, which is significantly higher than the previously reported models.

4.7.2.3   Prediction of Acute Inhalation Toxicity of Fluorocarbon Insulating Gases

Ye et al. [44] used the multiclass ARKA framework to develop models predicting acute inhalation toxicity of fluorocarbon insulating gases. Although the full form of ARKA has been incorrectly written in the publication, the authors showed that the

Gradient Boosting model trained using 16 ARKA descriptors outperformed 30 descriptor Support Vector Machine and Gradient Boosting QSAR models, both in terms of cross-validation score and in terms of prediction errors.

4.8 Conclusion

Activity cliff identification and analysis have remained a challenge for cheminformaticians worldwide, due to their non-compliance with the similarity principle. In this realm, the ARKA framework has provided a simple and efficient cheminformatic tool for the identification of activity cliffs, confident and less-confident data points, and a framework for the assessment of the modelability of a dataset. Moreover, as a supervised dimensionality reduction technique, the ARKA and multiclass ARKA descriptors showed improved model performances, although dimensionality reduction techniques are always associated with the loss of chemical information. The exponential increase in the number of applications of the ARKA and multiclass ARKA frameworks, since their introduction in 2024 and 2025, respectively, demonstrates their increased acceptability to the scientific community worldwide, paving the way for increased research on activity cliffs.

References

1. Stumpfe D, Hu H, Bajorath J (2019) Evolving concept of activity cliffs. ACS Omega 4:14360–14368. https://doi.org/10.1021/acsomega.9b02221
2. Guha R, Van Drie JH (2008) Structure−activity landscape index: identifying and quantifying activity cliffs. J Chem Inf Model 48:646–653. https://doi.org/10.1021/ci7004093
3. Golbraikh A, Muratov E, Fourches D, Tropsha A (2014) Data set modelability by QSAR. J Chem Inf Model 54:1–4. https://doi.org/10.1021/ci400572x
4. Ruiz IL, Gomez-Nieto MA (2018) Study of data set modelability: modelability, rivality, and weighted modelability indexes. J Chem Inf Model 58:1798–1814. https://doi.org/10.1021/acs.jcim.8b00188
5. Banerjee A, Roy K (2024) ARKA: a framework of dimensionality reduction for machine-learning classification modeling, risk assessment, and data gap-filling of sparse environmental toxicity data. Environ Sci Processes Impacts 26:991–1007. https://doi.org/10.1039/D4EM00173G
6. Wold S, Esbensen K, Geladi P (1987) Principal component analysis. Chemom Intell Lab Syst 2:37–52. https://doi.org/10.1016/0169-7439(87)80084-9
7. Zhou H, Wang F, Tao P (2018) t-Distributed stochastic neighbor embedding method with the least information loss for macromolecular simulations. J Chem Theory Comput 14:5499–5510. https://doi.org/10.1021/acs.jctc.8b00652
8. Crespo Márquez A (2022) The curse of dimensionality. In: Digital maintenance management. Springer series in reliability engineering. Springer, Cham. https://doi.org/10.1007/978-3-030-97660-6_7
9. Topliss JG, Costello RJ (1972) Chance correlations in structure-activity studies using multiple regression analysis. J Med Chem 15:1066–1068. https://doi.org/10.1021/jm00280a017

10. Varsou D-D, Banerjee A, Roy J, Roy K, Savvas G, Sarimveis H, Wyrzykowska E, Balicki M, Puzyn T, Melagraki G, Lynch I, Afantitis A (2024) The round-robin approach applied to nanoinformatics: consensus prediction of nanomaterials zeta potential. Beilstein J Nanotechnol 15:1536–1553. https://doi.org/10.3762/bjnano.15.121

11. Gajewicz A (2017) What if the number of nanotoxicity data is too small for developing predictive Nano-QSAR models? An alternative read-across based approach for filling data gaps. Nanoscale 9:8435–8448. https://doi.org/10.1039/C7NR02211E

12. Banerjee A, Kar S, Roy K, Patlewicz G, Charest N, Benfenati E, Cronin MTD (2024) Molecular similarity in chemical informatics and predictive toxicity modeling: from quantitative read-across (q-RA) to quantitative read-across structure–activity relationship (q-RASAR) with the application of machine learning. Crit Rev Toxicol 54:659–684. https://doi.org/10.1080/10408444.2024.2386260

13. Roy K, Kar S, Das RN (2015) Understanding the basics of QSAR for applications in pharmaceutical sciences and risk assessment. Academic Press, New York. https://doi.org/10.1016/C2014-0-00286-9

14. Snedecor GW, Cochran WG (1991) Statistical methods.8th edn. Wiley-Blackwell, Hoboken

15. Banerjee A, Roy K (2024) Machine learning assisted classification RASAR modeling for the nephrotoxicity potential of a curated set of orally active drugs. Sci Rep 15:808. https://doi.org/10.1038/s41598-024-85063-y

16. Banerjee A, Roy K (2024) The application of chemical similarity measures in an unconventional modeling framework c-RASAR along with dimensionality reduction techniques to a representative hepatotoxicity dataset. Sci Rep 14:20812. https://doi.org/10.1038/s41598-024-71892-4

17. Qin L-T, Zhang J-Y, Nong Q-Y, Xu X-C-L, Zeng H-H, Liang Y-P, Mo L-Y (2024) Classification and regression machine learning models for predicting the combined toxicity and interactions of antibiotics and fungicides mixtures. Environ Pollut 360:124565. https://doi.org/10.1016/j.envpol.2024.124565

18. Shan R, Zhang R, Gao Y, Wang W, Zhu W, Xin L, Liu T, Wang Y, Cui P (2025) Evaluating ionic liquid toxicity with machine learning and structural similarity methods. Green Chem Eng 6:249–262. https://doi.org/10.1016/j.gce.2024.08.008

19. Borah GS, Nagamani S (2024) Development of a robust machine learning model for Ames test outcome prediction. Chem Phys Lett 856:141663. https://doi.org/10.1016/j.cplett.2024.141663

20. Kar S, Gallagher A (2024) Comparative QSAR and q-RASAR modeling for aquatic toxicity of organic chemicals to three trout species: *O. Clarkii, S. Namaycush, and S. Fontinalis*. J Hazard Mater 480:136060. https://doi.org/10.1016/j.jhazmat.2024.136060

21. Rahimi-Soujeh Z, Safaie N, Moradi S, Abbod M, Sharifi R, Mojerlou S, Mokhtassi-Bidgoli A (2024) New binary mixtures of fungicides against Macrophomina phaseolina: machine learning-driven QSAR, read-across prediction, and molecular dynamics simulation. Chemosphere 366:143533. https://doi.org/10.1016/j.chemosphere.2024.143533

22. Abdellatif H, Laidi M, Si-moussa C, Amrane A, Euldji I, Benmouloud W (2025) Contributions to the development of prediction models for the toxicity of ionic liquids. Struct Chem 36:865–886. https://doi.org/10.1007/s11224-024-02411-4

23. Sun T, Wei C, Liu Y, Ren Y (2024) Explainable machine learning models for predicting the acute toxicity of pesticides to sheepshead minnow (*Cyprinodon variegatus*). Sci Total Environ 957:177399. https://doi.org/10.1016/j.scitotenv.2024.177399

24. Banjare P, Murmu A, Matore BM, Singh J, Papa E, Roy PP (2024) Unveiling the interspecies correlation and sensitivity factor analysis of rat and mouse acute oral toxicity of antimicrobial agents: first QSTR and QTTR modeling report. Toxicol Res 13:1–12. https://doi.org/10.1093/toxres/tfae191

25. Qin L-T, Tian X-F, Zhang J-Y, Liang Y-P, Zeng H-H, Mo L-Y (2024) Comprehensive machine learning-based models for predicting mixture toxicity of azole fungicides toward algae (*Auxenochlorella pyrenoidosa*). Environ Int 194:109162. https://doi.org/10.1016/j.envint.2024.109162

26. Yang X, Yang Y, Watson P, Liu H (2025) Development of quantitative structure property relationship models and tool for predicting the soil adsorption coefficient ($logK_{OC}$). Environ Pollut 368:125703. https://doi.org/10.1016/j.envpol.2025.125703

27. Li N, Chen Z, Zhang W, Li Y, Huang X, Li X (2025) Web server-based deep learning-driven predictive models for respiratory toxicity of environmental chemicals: mechanistic insights and interpretability. J Hazard Mater 489:137575. https://doi.org/10.1016/j.jhazmat.2025.137575

28. Gao Y, Qiu Y, Wan F, Cui S, Zhao Q, Zhao Y, Zhang D, Zhang C, Zhou J, Liu W, Zhuang S (2025) PBScreen: a server for the high-throughput screening of placental barrier–permeable contaminants based on multifusion deep learning. Environ Pollut 370:125858. https://doi.org/10.1016/j.envpol.2025.125858

29. Yin Z, Zhang M, Liu R, Cai Y (2025) Explainable machine learning models enhance prediction of PFAS bioactivity using quantitative molecular surface analysis-derived representation. Water Res 280:123500. https://doi.org/10.1016/j.watres.2025.123500

30. Sun Z, Huo D, Guo J, Yan A (2025) Modeling and interpretability study of the structure–activity relationship for multigeneration EGFR inhibitors. ACS Omega 10:11176–11187. https://doi.org/10.1021/acsomega.4c10464

31. Bhattacharyya P, Das S, Ojha PK (2025) Risk assessment of industrial chemicals towards salmon species amalgamating QSAR, q-RASAR, and ARKA framework. Toxicol Rep 14:102017. https://doi.org/10.1016/j.toxrep.2025.102017

32. Yu X (2025) Predicting chemical toxicity towards *Raphidocelis subcapitata* with quantum chemical descriptors. Algal Res 89:104055. https://doi.org/10.1016/j.algal.2025.104055

33. Hao Y, Duan Z, Liu L, Xue Q, Pan W, Liu X, Zhang A, Fu J (2025) Development of an interpretable machine learning model for neurotoxicity prediction of environmentally related compounds. Environ Sci Technol 59:11108–11120. https://doi.org/10.1021/acs.est.5c03311

34. Wei J, Duan L, Zhao J, Tian L, He M (2025) QSAR model development for the environmental risk limits and high-risk list identification of phenylurea herbicides in aquatic environments. J Agric Food Chem 73:14867–14881. https://doi.org/10.1021/acs.jafc.5c01253

35. Adhab AH, Mahdi AS, Shukla M, Yadav A, Manjunatha R, Kumar S, Shit D, Sangwan G, Mansoor AS, Radi UK, Abd NS (2025) Estimation of activity coefficient of aqueous ionic liquids using a machine learning method: the artificial neural network coupled with group contribution approach. J Indian Chem Soc 102:101924. https://doi.org/10.1016/j.jics.2025.101924

36. Fang H, Li W, Lu X, Chen J, Zhang P, Qi S, Ge R-S, Wang Y (2025) Structure-activity relationships of halogenated disinfectants as potent inhibitors of 5α-reductase type 1: potential impact on neurosteroid synthesis. J Hazard Mater 496:139498. https://doi.org/10.1016/j.jhazmat.2025.139498

37. Banerjee A, Roy K (2022) First report of q-RASAR modeling toward an approach of easy interpretability and efficient transferability. Mol Divers 26:2847–2862. https://doi.org/10.1007/s11030-022-10478-6

38. Banerjee A, Roy K (2024) How to correctly develop q-RASAR models for predictive cheminformatics. Expert Opin Drug Discov 19:1017–1022. https://doi.org/10.1080/17460441.2024.2376651

39. Roy K, Banerjee A (2024) q-RASAR. A path to predictive cheminformatics. Springer, New York. https://doi.org/10.1007/978-3-031-52057-0

40. Sobanska AW, Banerjee A, Roy K (2024) Organic sunscreens and their products of degradation in biotic and abiotic conditions—in silico studies of drug-likeness and human placental transport. Int J Mol Sci 25:12373. https://doi.org/10.3390/ijms252212373

41. Banerjee A, Roy K (2025) The multiclass ARKA framework for developing improved q-RASAR models for environmental toxicity endpoints. Environ Sci Processes Impacts 27:1229–1243. https://doi.org/10.1039/D5EM00068H

42. Banerjee A, Roy K (2025) A new approach methodology (NAM) for carcinogenicity prediction of organic chemicals using the multiclass ARKA framework and machine-learning-based stacking regression. J Hazard Mater 496:139302. https://doi.org/10.1016/j.jhazmat.2025.139302

43. Baran K, Kloskowski A (2025) Unlocking structural insights into CO_2 absorption with ionic liquids: a comparative study of machine learning, association rules, and meta-learning for modeling Henry's law constant. ACS Sustain Chem Eng 13:12805–12817. https://doi.org/10.1021/acssuschemeng.5c06122
44. Ye F, Xiao F, Zhan A, Chu Y, Tian S, Zhang X (2025) QSAR-based prediction of acute inhalation toxicity and SHAP interpretability analysis of fluorocarbon environmental-friendly insulating gases. Environ Res 285:122340. https://doi.org/10.1016/j.envres.2025.122340

Chapter 5
Activity Cliffs and Dataset Modelability: Future Roadmaps

Abstract The similarity levels among different compounds are based on the considered feature set and the measures used to define the similarity. A comprehensive analysis of the effects of features and similarity definitions on identifying activity cliffs and the resultant impact on the data set modelability is warranted. The knowledge of activity cliffs may also be exploited to explore better the structural features responsible for compound potencies across different target sets. Further application of machine learning methods in the identification of activity cliffs is also recommended.

Keywords Activity cliffs · Dataset modelability · Structural features · Machine learning

The level of similarity computed among different compounds depends on the underlying features used to define the similarity and also the similarity function being used. The identification of appropriate chemical features is very important for similarity analysis in the process of identifying activity cliffs. The concepts of read-across and quantitative read-across structure-activity relationship (q-RASAR) [1] are relevant in the context of explaining the activity-cliff behavior of chemical compounds. Exploration of an appropriate similarity function may also be an interesting area of study.

Although the concepts of the SALI index [2], SAR index [3], and SAS map [4] are not new in medicinal chemistry for activity cliff determination, their use is less common in the literature compared to the identification of the applicability domain for QSAR models. The presence of activity cliffs in the modeling data sets directly influences their modelability. Despite several isolated reports on the modelability [5] and rivality [6] indices, these approaches are not routinely used in quantitative structure-activity/property relationship (QSAR/QSPR) studies. The recent introduction of the Arithmetic Residuals in K-Groups Analysis (ARKA) approach [7] for activity cliff identification has found several applications in QSAR/QSPR modeling for drug development, environmental toxicology, and materials science. The ARKA

K. Roy, A. Banerjee, *Activity Cliffs*, SpringerBriefs in Molecular Science, https://doi.org/10.1007/978-3-032-10081-8_5

descriptor algorithm mimics the most discriminating feature selection strategy [8] for positive and negative classes; however, alternative approaches, possibly incorporating machine learning techniques, may also be explored. Besides activity cliffs, this approach also indicates less confident and less modelable data points, which can help explain poor model quality or inaccurate predictions. The distance of a data point from the origin in the ARKA plot also serves as a confidence measure, considering its location in a specific quadrant and the experimental endpoint value. Along with ARKA values, analyzing close congeners through similarity analysis, relevant in read-across or q-RASAR techniques, can effectively explain activity cliff behavior. Using an appropriate visualization method, such as Uniform Manifold Approximation and Projection (UMAP) [9] for exploring the chemical space, can complement the findings from ARKA analysis.

The introduction of multiclass ARKA [10] demonstrates the dependence of response on the descriptors in a response range-specific manner. This means that a particular descriptor may not have a similar influence on the response in different Y-ranges. This concept may further facilitate the identification of activity cliffs. While in traditional QSAR analysis, the average contribution of a particular descriptor over the extended response range is considered for modeling analysis, the range-specific contribution of descriptors provides additional insight into the identification of activity cliffs. Coupled with the q-RASAR technique, multi-class ARKA models can provide enhanced internal and external validation statistics compared to the traditional QSAR models. Only a limited number of studies have been conducted on the modified q-RASAR technique involving multi-class ARKA; extensive applications of this approach are needed to cover different areas of medicinal chemistry and predictive toxicology problems.

The failure of QSAR models may not always be due to statistical reasons, as proper identification of cliffs in an otherwise smooth structure-activity landscape may be the starting point of considering a distinct mechanism of action for a new series of congeners. The removal of activity cliffs may be critical to the performance of methodologically sound QSAR models [11]. When identifying activity cliffs, the quality of experimental data used for decision-making is crucial. Data for different congeners should be obtained uniformly using the same experimental protocol; otherwise, the analysis of structure-activity relationships may be flawed. A recent approach involves identifying target set-dependent activity cliffs by considering differences in potency values and determining set-dependent potency difference thresholds [12]. These considerations account for structure-activity relationships within a specific compound class. Besides 2D-descriptors and fingerprint-based similarity, 3D-structure-based similarity information can offer valuable case studies in medicinal chemistry when exploring structure-activity landscapes. Finally, QSAR model development is not a push-button exercise [13]. A judicious analysis of the modelability of the data sets and the identification of activity cliffs, possibly using increasingly more advanced machine learning methods [14], can lead to a more satisfying quality of data gap filling.

References

1. Banerjee A, Kar S, Roy K, Patlewicz G, Charest N, Benfenati E, Cronin MTD (2024) Molecular similarity in chemical informatics and predictive toxicity modeling: from quantitative read-across (q-RA) to quantitative read-across structure–activity relationship (q-RASAR) with the application of machine learning. Crit Rev Toxicol 54:659–684. https://doi.org/10.1080/10408444.2024.2386260

2. Guha R (2012) Exploring structure–activity data using the landscape paradigm. WIREs Comput Mol Sci 2:829–841. https://doi.org/10.1002/wcms.1087

3. Peltason L, Bajorath J (2007) SAR index: quantifying the nature of structure−activity relationships. J Med Chem 50:5571–5578. https://doi.org/10.1021/jm0705713

4. Bajorath J, Peltason L, Wawer M, Guha R, Lajiness MS, Van Drie JH (2009) Navigating structure–activity landscapes. Drug Discov Today 14:698–705. https://doi.org/10.1016/j.drudis.2009.04.003

5. Golbraikh A, Muratov E, Fourches D, Tropsha A (2013) Data set modelability by QSAR. J Chem Inf Model 54:1–4. https://doi.org/10.1021/ci400572x

6. Ruiz IL, Gómez-Nieto MA (2019) Building of robust and interpretable QSAR classification models by means of the rivality index. J Chem Inf Model 59:2785–2804. https://doi.org/10.1021/acs.jcim.9b00264

7. Banerjee A, Roy K (2024) ARKA: a framework of dimensionality reduction for machine-learning classification modeling, risk assessment, and data gap-filling of sparse environmental toxicity data. Environ Sci Process Impacts 26:991–1007. https://doi.org/10.1039/D4EM00173G

8. Murcia-Soler M, Perez-Gimenez F, Garcia-March FJ, Salabert-Salvador MT, Diaz-Villanueva W, Castro-Bleda MJ, Villanueva-Pareja A (2004) Artificial neural networks and linear discriminant analysis: a valuable combination in the selection of new antibacterial compounds. J Chem Inf Comput Sci 44:1031–1041. https://doi.org/10.1021/ci030340e

9. McInnes L, Healy J, Saul N, Großberger L (2018) UMAP: uniform manifold approximation and projection. J Open Source Softw 3:861 https://doi.org/10.21105/joss.00861

10. Banerjee A, Roy K (2025) The multiclass ARKA framework for developing improved q-RASAR models for environmental toxicity endpoints. Environ Sci Process Impacts 27:1229–1243. https://doi.org/10.1039/D5EM00068H

11. Maggiora GM (2006) On outliers and activity cliffs—why QSAR often disappoints. J Chem Inf Model 46:1535. https://doi.org/10.1021/ci060117s

12. Stumpfe D, Hu H, Bajorath J (2019) Evolving concept of activity cliffs. ACS Omega 4:14360–14368. https://doi.org/10.1021/acsomega.9b02221

13. Gramatica P, Cassani S, Roy PP, Kovarich S, Yap CW, Papa E (2012) QSAR modeling is not "push a button and find a correlation": a case study of toxicity of (benzo-)triazoles on algae. Mol Inform 31:817–835. https://doi.org/10.1002/minf.201200075

14. Banerjee A, Roy K, Gramatica P (2025) A bibliometric analysis of the cheminformatics/QSAR literature (2000–2023) for predictive modeling in data science using the SCOPUS database. Mol Divers 29:3703–3715. https://doi.org/10.1007/s11030-024-11056-8

Index

If you have any concerns about our products,
you can contact us on
ProductSafety@springernature.com

In case Publisher is established outside the EU,
the EU authorized representative is:
Springer Nature Customer Service Center GmbH
Europaplatz 3, 69115 Heidelberg, Germany

Printed by Libri Plureos GmbH
in Hamburg, Germany